# Introduction to Geographic Field Methods and Techniques

JOHN F. LOUNSBURY
AND
FRANK T. ALDRICH

Arizona State University

# Introduction to Geographic Field Methods and Techniques

## SECOND EDITION

Macmillan Publishing Company
New York

Maxwell Macmillan Canada
Toronto

Maxwell Macmillan International
New York   Oxford   Singapore   Sydney

This book was set in Helvetica.
Cover Design Coordination: Cathy Watterson
Production Coordination: Anne Daly
Cover photo courtesy of United States Department of Agriculture

**Macmillan Publishing Company**
**113 Sylvan Avenue, Englewood Cliffs, NJ 07632**

Library of Congress Catalog Card Number: 85-072217
International Standard Book Number: 0-675-20509-3
Printed in the United States of America
      5  6  7  8  9—91  92  93  94

Alternative:

Library of Congress Catalog Card Number: 85-072217
International Standard Book Number: 0-675-20509-3
Printed in the United States of America
      5  6  7  8  9—94  93  92  91

# Contents

# Preface

Direct observation as a major source of information is deeply ingrained in the history of geography. Field studies have long served both as a research tool and as a method of teaching. However, a great deal of significant geographical research has been done in recent years that is not, nor should be, based on data obtained directly in the field. The nature of this research is such that primary sources of data are of little or no benefit. On the other hand, there are geographical research problems that depend, in part or in total, on information acquired directly from field observations. In such cases, critical data are not available from published sources, or, if available, do not exist at the scale or date desired, or for other reasons are unreliable or unsuitable for the purposes of a specific problem.

This book presents several approaches and a range of methods and techniques concerned with the systematic acquisition of data in the field. It is primarily for those who aspire to become applied geographers and anticipate that field study will be an important aspect of their future research and that field training is an essential part of their overall formal training. It is general rather than specific in design, and it is written for students and researchers in geography and related fields who have little or no field experience.

At the present, about 75 percent of the geography departments offer courses that cover field techniques. These courses are of two major types: (1)

courses where the major emphasis is on the instruction in field techniques and a variety of field methods, techniques, and technical skills pertinent to a range of subject matter are covered; and (2) courses where some field techniques are included, but in which the focus is on a specific topic such as urban geography, geomorphology, and so forth. This book is designed primarily for courses of the first type in that it includes subject matter pertinent to most of the major subfields in geography, both physical and cultural. It does not address itself to highly sophisticated field techniques relevant to advanced research in a specific area of specialization. It is not a textbook on surveying, physical geography, or applied urban geography, etc.

Emphasis is placed on the methods and techniques used to obtain primary data from field observations, and the content is not directly concerned with map or photo interpretation, statistical techniques, methods of analysis, or theory common to other areas of research. It is assumed that the researcher will apply pertinent concepts, knowledge of subject matter, and skills acquired in other courses and experiences to the solution of the specific field problem undertaken.

This is not, nor necessarily should there be, a "standard" geography field course. Areas, institutions, student bodies, and interests and backgrounds of instructors are different. The effective field course must be adapted to the local situation. For this reason, this book is not designed for a particular course, but rather provides a framework within which the instructor can delete or elaborate certain sections to be compatible with the overall course objectives. Finally, field research is learning by doing. There is no substitute for actual field experience. This book provides the theory, philosophy, and conceptual structure for a variety of field techniques, but the student becomes proficient in field research only by doing field work.

The authors have instructed field courses and have been engaged in field research continuously for forty years and twenty years respectively. Together they have team taught a geographic field course at least once a year for seventeen years. The methodologies and techniques in this book have been the ones they have found to be most appropriate and workable for an introductory course. New material has been added throughout this new edition but particularly in the section dealing with base maps (Chapter 2) to include the use of orthophotoquadrangles and satellite imagery. Chapter 3, dealing with instrumentation, has been expanded significantly to include information on microcomputers and experiential field work. A number of new examples relating to sampling methods and questionnaires, including cognitive regional data, have been added to Chapters 4 and 5. Chapter 6 has a new section dealing with proposal construction. Two new appendices deal with land classification systems and measurement scales. Existing appendices have been updated and expanded.

The second edition required the efforts and cooperation of many individuals. The authors wish to especially acknowledge Kathy Nee of Charles E. Merrill Publishing Company for her sound advice and encouragement on many occasions. They also wish to thank Anne Daly for her superior editorial work. Thanks too, go to Laura Conkey of Dartmouth College, John Westfall of San Francisco State University, Sona Andrews of the University of Minnesota, and Lawrence Grossman of the Virginia Polytechnic Institute for reviewing the manuscript and making constructive suggestions. Finally, the authors wish to acknowledge the several hundred graduate and undergraduate students with whom they have had contact in field courses over the years. We hope that they have learned as much from us as we have learned from them.

J. F. LOUNSBURY
F. T. ALDRICH

# Foreword

Fieldwork has an ancient and honorable tradition in Geography. The best way to get to know a place is to see it first-hand, and no one can claim that he truly knows an area until he has seen it for himself. "The geographer and the geographer-to-be," said Carl O. Sauer, "are travelers, vicarious when they must be, actual when they may." The very first geographers were travelers who enthralled their friends with stories of the marvels and wonders they had seen in other lands, and even today some of us are not above inflicting tall tales from our travels on anyone who is willing to sit still long enough to listen to us.

The casual traveler may get the sense of an area, the "feel" of it, especially if he is a sensitive observer, but the observations of travelers often are subjective, impressionistic, and unsystematic. As Geography began to mature its practitioners realized that knowledge of and insight into the nature of places and areas must be based on more than the fleeting glimpses of a passing traveler, and they began to develop techniques of more careful observation, more precise measurement, and more accurate recording of the features in which they were interested. The development of techniques for the systematic acquisition of otherwise unavailable data came to occupy an important, perhaps too important, role in Geography before World War II. Some few even went so far as to canonize "fieldwork," and they argued that a geographer had not really won his spurs until he had gone out and gotten his boots dirty.

Dirty boots have largely gone out of fashion in Geography since World War II. The focus of the discipline has shifted, some would say too far, from the empirical and the concrete to the abstract and theoretical. Many geographers have become fascinated with the vistas that have been opened to them by the wizardry of the Age of Electronics, and the ever wider range and variety of data collected and published by public agencies have enabled geographers to rely more and more on secondary sources of information.

Even today, however, geographers often are compelled to go into the field to collect data that otherwise are not available. First-hand data collection can be a salutary educational experience for those who have relied too heavily and uncritically on secondary data, because all too often they have failed to appreciate the problems and difficulties of collecting data in the field, and they have remained blissfully unaware of the flaws and errors that inevitably result from the arbitrary decisions in classification that must be made when attempting to do so. Fieldwork should breed a healthy skepticism of all published data.

Fieldwork is a vigorous activity that demands the fullest exercise of all one's talents, energies, and abilities, mental as well as physical. It provides the pleasure of learning that there is much good reading on the land for those who develop the skills needed to decipher and enjoy it, and few experiences can equal the thrill of personal discovery. Every venture in field research makes its own demands and poses its own problems, and there are as many field techniques as there are unanswered questions and geographers to tackle them; the only limit is the ingenuity of the individual scholar. The character and content of field study and field courses in Geography vary enormously from department to department and from instructor to instructor, and no one should attempt to prescribe a standard field course, yet certain standard techniques and procedures have been honed through time. The authors of this volume have wisely refrained from trying to tell anyone else how to conduct a field course; instead they have served up a veritable smorgasbord of the methods that geographers have developed to facilitate the collection of information in the raw.

JOHN FRASER HART

# Nature, Scope, and Objectives of Geographic Field Methods and Techniques

## Chapter

# 1

# Chapter 1

The terms *geographic field methods* and *techniques* are not precisely defined to the satisfaction of all professional geographers. The common usages of these terms cover a surprisingly large range of activities. For the purpose of this text, geographic field methods and techniques refer to the systematic acquisition of new or raw data within a specific research area. This includes an organized recording of observations made in the field within a defined spatial matrix or research area and the utilization of systems of data classification subject to subsequent processing, presentation, and analysis. The term *methods* is used to describe the overall research framework or design, and *techniques* refers to the actual manner in which field data are collected. Although these definitions are restraining, they will serve to establish the limitations of the subject matter to be discussed.

## THE NATURE OF MODERN GEOGRAPHY

Geography is a broad discipline and focuses not on one particular set of phenomena, but on spatial distributions of many phenomena and their areal associations as they occur on the surface of the earth. As a result, geography consists of several well-established and diverse subfields. But although well-defined and generally accepted unifying threads or themes unite the discipline as a whole, the large number and variety of subfields is often a puzzle to nongeographers. Indeed it has been a subject of serious discussion among professional geographers for decades. There are almost sixty recognized topical or systematic subfields or areas of specialization and over sixty major areal or regional specializations (see Table 1-1).

Subfields of geography may be categorized as either regional or topical. Regional or areal geography considers subject matter of several topical or systematic subfields as it pertains to a given area or region. A topical or systematic subfield, on the other hand, normally focuses on one set of phenomena wherever it may occur, perhaps in many areas or regions. Manufacturing, climate, settlement types, landforms, cities, and vegetation are but a few examples of topical subfields. Field methods, cartography, quantitative methods, remote sensing, and electronic data processing are considered topical subfields too, but they focus on methods and techniques of obtaining, analyzing, or processing information that may be derived from several other subfields. Similarly subfields like educational geography and environmental protection draw on subject matter from other subfields. Historical geography concentrates on events and things in the past effecting the spatial distribution of phenomena in the present. The differences between regional and topical geography are basically differences of approach rather than differences of subject matter. They often overlap or merge in practical study or research.

---

## TABLE 1-1

*Major Subfields of Geography*

### Topical Proficiency Codes

| | | | |
|---|---|---|---|
| 01 | Administration | 29 | Planning, Urban |
| 02 | Agricultural Geography | 31 | Political Geography |
| 03 | Economic Development | 32 | Population Geography |
| 04 | Applied Geography | 33 | Recreational Geography |
| 05 | Arid Regions Geography | 34 | Regional Geography |
| 06 | Biogeography | 35 | Resource Geography |
| 07 | Climatology | 36 | Rural Geography |
| 08 | Coastal Geography | 37 | Soils Geography |
| 09 | Environmental Studies (Conservation) | 38 | Transportation & Communication |
| 10 | Cultural Geography | 39 | Tropical Geography |
| 11 | Economic Geography | 40 | Urban Geography |
| 12 | Educational Geography | 41 | Water Resources |
| 13 | Location Theory | 42 | Field Methods |
| 14 | Geomorphology | 43 | Land Use |
| 15 | Historical Geography | 44 | Quantitative Methods |
| 16 | History of Geography | 46 | Teaching Techniques |
| 17 | Marine Resources | 47 | Audio-Visual Materials & Techniques |
| 19 | Manufacturing Geography | 48 | Remote Sensing |
| 20 | Marketing Geography | 49 | Social Geography |
| 21 | Medical Geography | 52 | Cartography, General |
| 23 | Military Geography | 55 | Librarianship, Geographical |
| 25 | Oceanography | 56 | Electronic Data Processing |
| 27 | Physical Geography | 57 | Environmental Perception |
| 28 | Planning, Regional | | |

## TABLE 1-1 (continued)

### Areal Proficiency Codes

| | | | |
|---|---|---|---|
| 01 | World | 32 | West Indies & Bermuda |
| 06 | North Polar Region | 33 | South America |
| 07 | South Polar Region | 34 | Northern South America & West Coast |
| 08 | North America | 35 | Brazil |
| 09 | Anglo-America | 36 | Southern South America |
| 10 | Canada | 37 | Europe |
| 11 | Territories | 38 | Scandinavia |
| 12 | British Columbia | 39 | Western Europe |
| 13 | Prairie Provinces | 40 | British Isles |
| 14 | Ontario | 41 | Southern Europe |
| 15 | Quebec | 42 | Central Europe |
| 16 | Maritime Provinces | 43 | Eastern Europe |
| 17 | United States of America | 44 | USSR |
| | AAG Division Boundaries: | 45 | Asia |
| 18 | New England | 47 | China |
| 19 | Middle States (N.Y., N.J., Del., E. Pa.) | 48 | Japan |
| 20 | Middle Atlantic (Wash. D.C., Baltimore | 49 | Southeast Asia |
| | metro area) | 50 | South Asia |
| 21 | Southeastern | 53 | Southwest Asia (Middle East) |
| 22 | Southwestern | 54 | Africa |
| 23 | Pacific Coast | 55 | North Africa |
| 24 | Great Plains-Rocky Mountains | 56 | West Africa |
| 25 | West Lakes | 57 | Central Africa |
| 26 | East Lakes | 58 | East Africa |
| 28 | Latin America (Hispanic America) | 59 | Africa South of Congo |
| 29 | Middle America (Caribbean) | 60 | Australia |
| 30 | Mexico | 61 | New Zealand |
| 31 | Central America | 62 | Pacific Islands |

### Specialty Group Codes

| | | | | | |
|---|---|---|---|---|---|
| 01 | Africa | 14 | Environmental Perception | 24 | Native American |
| 02 | Aging | 15 | Environmental Studies | 25 | Political Geography |
| 03 | Applied | 16 | Geographic Perspectives | 26 | Population |
| 04 | Asian | | on Women | 28 | Regional Development & |
| 05 | Bible | 17 | Geography in Higher | | Planning |
| 06 | Biogeography | | Education | 27 | Recreation |
| 07 | Canadian Geography | 18 | Geomorphology | 29 | Remote Sensing |
| 08 | Cartography | 19 | Historical | 30 | Rural Development |
| 09 | Chinese Geography | 20 | Industrial Geography | 31 | Socialist |
| 10 | Climatology | 21 | Latin American | 32 | Soviet |
| 11 | Coastal & Marine | 22 | Mathematical Models & | 33 | Transportation |
| 12 | Cultural Ecology | | Quantitative Methods | 34 | Urban |
| 13 | Energy | 23 | Medical | 35 | Water Resources |

The Association of American Geographers recognizes fifty-seven major topical or systematic specializations and sixty-two major areal or regional subfields. In addition, the AAG has established thirty-five "Specialty Groups" that act as mechanisms for geographers with similar backgrounds, training, and interests to exchange information and investigate problems of mutual concern.

A second apparent dichotomy in geography concerns the division of the discipline into physical or human geography depending upon the major phenomena studied. This division is not unique to geography; anthropology and psychology also study topics that are either physical or cultural in nature. Physical geography focuses primarily on the natural sciences and deals with the atmosphere, hydrosphere, biosphere, and lithosphere. Climatology, geomorphology, biogeography, and oceanography exemplify subfields under the broad umbrella of physical geography. Human geography, on the other hand, studies economic, cultural, social, and political subjects. Population geography, medical geography, manufacturing geography, and economic development are but a few of the human geography subfields. As with the division between topical and regional geography, the distinction between physical and human geography is a convenient division on paper, but in practice (especially in field research) the fields frequently overlap.

Field geographers today study phenomena as different as glacial landforms in Michigan and house types in Papua, New Guinea. But although the data they collect and the field research techniques they employ may vary, the overall research design and the emphasis on spatial distributions will be similar.

Furthermore, each field problem is unique, and the research design and specific techniques used to acquire data must be adjusted to serve its needs. Each field problem has its special objectives, its particular research area, and its own set of essential data to be acquired. There is no overall master design or standard set of techniques that can be applied to the solution of all field problems. A significant part of field work is the intellectual exercise of devising a research format to fit the needs of a specific problem. To be sure, there are guidelines, such as methodologies and techniques, that have proved successful in the completion of other field problems. However, these methods and techniques must be carefully studied and adapted in their present form, or modified or discarded, as the case may be.


## THE ROLE OF FIELD METHODS AND TECHNIQUES IN MODERN GEOGRAPHICAL RESEARCH

A great deal of contemporary geographical research is not, nor need be, based on data acquired in the field. Such research may use secondary or tertiary sources of information or may be concerned with problems of a theoretical nature or of a scale or level of resolution for which primary sources of data are of little or no benefit. Detailed field research is often costly in terms of both time and money and should be undertaken only when critical information for the solution of a given problem cannot be obtained in any other manner.

There are, however, certain geographical research problems where essential data are not available from published sources or the existing data are outdated or otherwise unreliable. Such problems occur more often in applied than in theoretical research. Research problems that frequently require field work may be classified into highly generalized groups as follows:

1. Microscale problems in which the data required are so detailed that no published sources exist and for which the data cannot be obtained from map, aerial-photograph, or other remote-sensing imagery interpretation.
2. Problems that are concerned with dynamic areas which undergo changes over very short periods of time.
3. Problems that require information that does not reflect itself as visible features of the landscape, such as certain types of social and economic behavioral patterns (attitudes, perceptions, shopping behavior, where people go for medical care, where crops or manufactured items are marketed, etc).

Some new research thrusts of the last two decades concerned with generalizations, theories, and laws in geography do not require field work as we have defined it; therefore it may be stated that field study is not essential to the training of all professional geographers. However, for those professional geographers and potential professional geographers who address themselves to applied problems concerned with land use, environmental quality, planning, and so forth, field work is a necessary part of their overall training. Today, more scientists in various fields are working on the pressing domestic problems facing the nation. Many of these problems focus on applied research and many are concerned with problems at the intermediate and microscale. Many applied geographers are included in this group and much of their work is field-oriented.

## HISTORICAL DEVELOPMENT OF FIELD INVESTIGATIONS

The nature and scope of modern field research have developed over a long period of time. Direct observation as a major source of information is deeply entrenched in the history of geography. Field studies have long served both as a research tool and a method of teaching. A century ago, prestigious geographers visited distant lands and returned to report their impressions and general observations to the professional societies of the day. Although highly unscientific by modern standards, these reports were important contributions because they added new information to the pool of knowledge. As is true in other sciences, the techniques and equipment used by field geographers have evolved over the years. Direct observation is still important in many modern research problems, but today these observations can be recorded accurately, organized systematically, and analyzed.

Before the mid-1920s, field work in geography consisted of random observations and impressions which varied greatly from one researcher to another. Field studies were highly local in scope and impact, and high-level accuracy was rare. Suitable base maps were not available for most areas, quantitative field techniques had yet to be developed, and no overall methodology, structure, or framework existed. Beginning in the late 1920s, several important developments took place that established the foundations for modern field work and accelerated the evolution of field investigations. In the 1920s and 1930s, large-scale topographic maps became available for extensive areas of the country. The detailed scale and abundance of orientation points on these topographic maps, or quadrangles, provided the field researcher with suitable base maps for detailed and accurate field work. The maps themselves provided a large body of information to supplement actual field observations.

Shortly after World War I, diverse technological breakthroughs—cameras, photographic developing processes, modern types of aircraft, etc.—allowed for the production of a *vertical aerial photograph* suitable for precision field mapping; and by the 1930s, aerial photographic coverage for much of the country was available. The aerial photograph, with its abundance of orientation or control points, provided an excellent base map and a wealth of data that did not have to be recorded in the field. It enabled the field researcher to portray actuality with a high degree of accuracy and at a speed of operation that allowed large areas to be surveyed. In recent years, a wide variety of additional *remotely sensed imagery*, taken from high-level aircraft and satellites, has been developed. Much of this imagery provides data not visible to humans. However, these data can be translated into visible color or black-and-white images.

Sophisticated methods of field techniques and sampling procedures were also developed beginning in the late 1920s. In 1925, the *fractional code system of mapping* was developed by a group of geographers, including the late Vernor C. Finch, Wellington D. Jones, and Derwent Whittlesey. This technique provided a mechanism whereby several features of the landscape, both physical and cultural, can be mapped simultaneously with accuracy and speed and made subject to quantitative analysis.

A discourse on the history of the development of modern geographic field methods and techniques would not be complete without the mention of three major field mapping programs: The Land Utilisation Survey of Britain, The Land Classification Program of the Tennessee Valley Authority, and The Rural Land Classification Program of Puerto Rico. Each of these programs took several years, covered thousands of square miles, involved the efforts of many geographers, and received widespread notice and acclaim. And each in its own way developed and refined techniques that contributed greatly to the development of modern field studies.

## The Land Utilisation
## Survey of Britain

The Land Utilisation Survey of Britain was established in the fall of 1930. The late L. Dudley Stamp, Professor of Geography, University of London, London School of Economics, organized the program and served as its director from the beginning. The major objective of the program was to make an inventory of the land resources of every acre in England, Wales, Scotland, and Isle of Man, relying mainly on direct observations in the field. Fortunately, Britain by this time had complete topographic map coverage at a detailed scale which included topography, drainage, field patterns, roads, and buildings. These maps, with the scale of 1:10,560 (6 inches = 1 mile; 1 centimeter = 105.6 meters), were used as the base maps for the field mapping. Each of the maps covered 6 square miles, and over 22,000 such maps were required to cover the entire research area. The actual field mapping took place during 1931–1934, and the analysis of the data and development of the land capability classifications took several years longer. The land-use data in conjunction with data on slope, soil, drainage, and other relevant factors served as the basis for the Classification of Land for Land-Use Planning, one of the first attempts at classifying land capability. Ten major types of land were defined, as well as the optimum use for each. The analysis of the total survey was published in ninety-two reports bound into nine volumes.

The Land Utilisation Survey of Britain was a major undertaking for its time and was a landmark in the development of field studies in several respects. It was the first program to cover an extensive area [over 88,000 square miles (227,920 square kilometers)], to standardize mapping techniques and classification systems to insure accuracy and compatibility of the field data from one area to another, and to establish a land-use plan for the optimum development of the land in Britain. The success of this program led to the formal establishment of a commission, under the auspices of the International Geographical Union, to investigate the possibilities of developing a world land-use survey, modeled in part after the British program. Due to the tremendous complexity and magnitude of the proposed project, only a few pilot studies were completed. Furthermore, such a project required a degree of international cooperation that has not existed in recent history.

### The Land Classification Program of the Tennessee Valley Authority (TVA)

The Land Classification Program of the TVA began in the early 1930s under the directorship of G. Donald Hudson, who later became Chairperson of the Department of Geography at Northwestern University, then at the University of Washington. This program was concerned with the inventory, appraisal, and land capabilities of the area under the jurisdiction of the TVA. The program developed an elaborate fractional code system of mapping as well as other methods of detailed field analysis and sampling. It was the first time that the fractional code method was employed on an extensive scale combined with vertical aerial photographs. Aerial mosaics at the scale of 1:24,000 (1 inch = 2000 feet; 1 centimeter = 240 meters) were used as base maps for field mapping with a minimum mapping unit of 200 acres (about 81 hectares).

Land use, quality of land use, field size, amount of idle land, quality of farmsteads and equipment, slope, drainage, erosion, stoniness, rock exposure, and soil depth and fertility were all recorded within the highly structured fractional code. From these data, a series of *land capabilities* were developed. Land was classified on the basis of agricultural quality of the existing physical characteristics of the land, the quality of the existing use of the land, and the economic status of the people and the physical conditions of the land (relative severity or absence of problems and needs for readjustments).

The work carried on by the TVA group of geographers was of major importance in the evolution of modern field methods. The use of a highly structured and elaborate fractional code system proved to be a practical method of collecting great numbers of data quickly, accurately, and subject to quantitative analysis. The fractional code, in all its forms, dealt with eighty-four separate items. The development, classification, and definition of these items were themselves major undertakings. For the first time, vertical aerial photographs were used to map a large area and proved to be superior base maps for detailed field analysis. The *unit area method* of land classification, as developed by the TVA group, presented a mechanism that achieved the greatest possible accuracy in field investigation within a practical framework of time and cost. It served as a model for future land-use studies in that it presented an innovative structure to quantify occupancy patterns, both the observable features of the landscape as well as the non-visible phenomena.

## The Rural Land Classification Program of Puerto Rico

The Rural Land Classification Program of Puerto Rico began in 1949. It was carried out under the overall direction of Clarence F. Jones, Professor of Geography, Northwestern University. G. Donald Hudson, then Chairperson of the Geography Department, Northwestern University, and Rafael Pico, then Chairperson of the Puerto Rico Planning Board, were highly instrumental in the organizational phase of the program. John F. Lounsbury, then a graduate student at Northwestern University, served as the Assistant Program Director during the operational and field survey phases of the program. The actual field mapping and data collecting took place during 1949–1951. The data analysis, in addition to special field studies, was carried on for several additional years.

The entire Island of Puerto Rico and adjacent small islands (3435 square miles, or 8896 square kilometers) were surveyed. The field mapping was done on vertical aerial photographs at the scale of 1:10,000 (1 inch = approximately 833 feet; 1 meter = 10 kilometers). A detailed fractional code system of mapping was developed, patterned to some degree after those used in the TVA studies. Specific items within the fractional codes included land use, land-use quality, cultural features, soil type, slope, drainage, erosion, and stoniness. The highly diverse physical nature of the Commonwealth of Puerto Rico gives rise to a great variety of soil types, land use, and occupancy patterns, which was reflected in the highly complex fractional codes that were employed. Concurrent with the field mapping were lengthy and formalized interviews, taken to reveal the nonobservable aspects of the area, such as farming practices (use of fertilizer, crop rotation systems, source of water, farm size, and so on), marketing of commodities, manufacturing procedures, etc. An elaborate, structured random sampling system was employed. Interviews were taken on one out of every 8 large farms, one in 16 parcelas, one in 32 sharecroppers, and one in every 64 rural dwellings. The area surveyed was divided into eighteen mapping units. The field work in each unit was done by a team of three: a team chief, a graduate student in training subsequently to become a team chief in four to six weeks, and an interviewer. The team chief and graduate student in training were advanced students in geography who met all the requirements for the Ph.D. degree except the completion and acceptance of the dissertation and who had had considerable field work experience. The interviewers were Puerto Ricans, chosen because of their previous professional experience in agriculture, interviewing, and familiarity with the area. Each interviewer had lived or worked in the mapping area to which he was assigned. Twenty advanced geography graduate students completed the field work, each working on a full-time basis for eight to eighteen months. Doctoral students from twelve universities were involved. About a dozen additional professional geographers were associated with the program on a part-time or consultant basis.

The contributions of the program were significant. In terms of benefit to the Commonwealth of Puerto Rico, the data served as the basis for many island-wide planning programs such as economic capability classifications, recommended land uses, planning of rural roads and highways, farm management, rural zoning, and the location of new manufacturing plants. The program was a significant part of the Commonwealth's overall "Operation Bootstrap." The twenty Ph.D. dissertations in geography written based on the survey data added greatly to the body of knowledge about Puerto Rico and tropical environments. The scope and magnitude of the field work were unmatched elsewhere in the tropics. The validity of vertical photographs and the fractional code system of mapping were substantiated, as was their flexibility to relate diverse environments. It was the first time in geographic field work that a highly structured questionnaire, administrated through the use of a systematic sampling procedure, was employed on an extensive scale.

Numerous other highly significant field projects and programs have influenced the development of field investigations. However, the three programs discussed here, due primarily to their time in history, and due to their scope in terms of the size of the areas covered, time span, diversity of data collected (physical, economic and cultural), and involvement of people, were studied widely and exerted a major influence during the years when the foundations of modern field work were established.

## THE SCOPE OF MODERN
## FIELD RESEARCH

In recent years, a tremendous number of field projects at the local, state, and regional level have been and are being carried on. These projects most frequently focus on pressing areal or regional problems. As such, their scope is less comprehensive than those historical programs previously mentioned. They may be concerned with the collection of specific data such as microclimatological information or perceptions and attitudes of people concerning a particular issue, or they may be multidimensional in scope, dealing with a wide diversity of data such as the total change of the environment in a highly dynamic area.

Many current field problems are concerned with the location of new manufacturing plants, shopping centers, parks and recreational facilities, or power installations. Detailed information about the physical characteristics of the land and socioeconomic factors must be collected and analyzed to determine the optimum sites if a given proposed facility is to thrive in and be compatible with the overall area. Related field projects focus on land-use developments and land-use planning. The expansion of urban land uses in agricultural and rural areas is a common problem in many sections of the country, and data collected in the field serve as the basis of planning and often minimize the conflicts arising from incompatible land-use developments.

In recent years many contemporary field studies have focused on hazards, both natural and man-made. No doubt this recent emphasis springs from the newly aroused public interest in existing and potential hazardous situations and the fact that little relevant data at an appropriate scale exists. Natural hazards include the development of land inappropriate to the slope of the land. For example, floodplains are periodically underwater, and steep slopes are subject to landslides and slips. Developing relevant ordinances and regulations to control building on floodplains and hillsides depends upon detailed field studies. Additional naturally hazardous situations relate to the depth and fluctuations of ground water tables, soil compaction, and the chemical composition and structure of the underlying bedrock. Recent disasters in many parts of the world resulting from volcanic and tectonic activities have spurred interest in detailed field studies of land use in these and adjoining areas as well as surveys of the attitudes of the people residing in potentially dangerous areas.

The explosive growth of urban areas and industrial complexes gives rise to a host of problems directly associated with human activities. The rapid increase in numbers and size of airports, for example, creates a set of problems such as crash hazards, air and noise pollution, and decline of property values. Similar difficulties develop as highways, railroads, and docking facilities, for example, expand beyond existing capacities and demand new transportational structures. Planning new airports, freeways, and other facilities requires detailed field inventories and analysis to understand their impacts on adjacent land. New technologies and the rapid expansion of urban ideas create serious problems of water and air pollution and the disposal of solid waste, radioactive materials, and chemical and metallurgical wastes. Hazards are not new to society, but the rapid urbanization and growth of technology coupled with concern about those hazards are relatively recent. The consequent need for new and accurate information at various scales has created a great demand for detailed field work and analysis.

The spatial distribution and intensity of perceptions and attitudes of the people regarding governmental actions such as services, zoning, proposed new tax changes, etc., can be obtained only by field study. The determination of perceptions is the focus of a great number of field projects in many parts of the country.

Today, the legal necessity of preparing environmental impact statements has drastically increased the number and intensity of field investigations. These field studies require the acquisition of economic, ecological, and social data not attempted previously. Many communities and areas today are concerned about growth exceeding existing resources such as water and energy. Questions dealing with the optimum size of population, whether or not growth limitations should be imposed, etc., require data that can only be acquired in the field. The development of new towns, the changing structure and function of the central business districts, mobility, and social-economic characteristics of the population are also widely studied.

Today, and in the foreseeable future, the need for pertinent information to define and resolve pressing problems is great. Much of this information is not available from published sources at the scale, date, and reliability required. It is most likely that the demand for primary data that can be obtained only through field studies will increase greatly in the near future.


## THE STRUCTURE OF MODERN FIELD RESEARCH

The organization of a field research problem differs from other research problems in only one respect. Research dealing with secondary and tertiary sources involves assessing and evaluating data but usually does not concern itself with the actual acquisition of new or raw data. A field research problem, on the other hand, requires that considerable additional effort be expended in planning, preparing, and finally collecting the essential raw data before any analysis can be undertaken. Careful prefield planning and preparation are necessary to insure that the information gathered is as accurate as possible and, at the same time, obtained as quickly as possible without jeopardizing its validity.

Like other research problems, field studies must be highly structured if the final results are to be of good quality. The formulation of the overall research design requires sound organization (See Figure 1-1). The major components of a field research study should include:

1. A clear statement of the problem,
2. Determination of the research area,
3. Formulation of hypotheses,
4. Identification of the necessary data,
5. Categorization or classification and scale of the data to be obtained,
6. Acquisition of the data,
7. Processing and analyzing the data, and
8. Formulation of the answers to the question or problem as first stated.

The first step in any research plan is to identify the problem. Since problems are essentially questions that have no satisfactory answers at the time, one way to define the research problem is to present it in the form of a question. Geographic problems deal with spatial arrangements of phenomena. The phenomena studied may be physical or cultural, concrete or abstract, or features or ideas. Geographic field problems require study in the field to obtain primary data pertinent to the unanswered question. A clear and concise statement of the problem is most important since the problem or question will determine the scope and nature of the entire research plan.

The second step is to identify the research area within which the field research will be conducted. There is no standard or generally accepted size of a research area. It may be large or very small depending upon the needs of a given problem. The size of the area, however, will bear directly on the types of phenomena that can be studied as well as on the manner in which they are classified and the scale in which they are observed and recorded. Specific and detailed data may be obtained within a small research area, while more general data are gathered from large areas. The size of the area will also determine whether full field coverage can be obtained or sampling procedures need to be employed. In some field

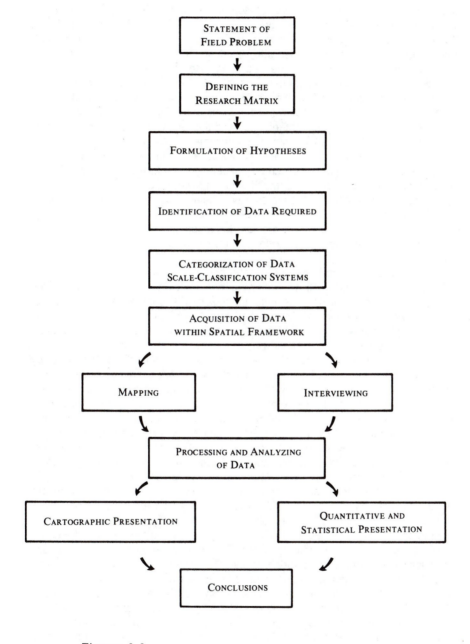

**Figure 1-1.**

The Components of a Field Research Problem.

research, the statement of the problem defines the area. For example, if the problem is stated, "What is the occupational structure of the gainfully employed in Town X?" or "What changes in land use have occurred in County Y between 1975 and 1980?", the area has been predetermined. On the other hand, if the problem is to determine how vegetation changes with altitude in a given climatic zone, the area is not predetermined and must be defined to include an area with significant topographical relief that a sufficient amount of relevant information may be obtained. The area provides the framework within which the field work will be carried out, and its precise areal extent must be determined to focus and further define the purposes of the research. The delimitation criteria used and the rationale for selecting a particular research area over all others should be documented clearly.

Most field research problems have one or more stated hypotheses. An *hypothesis* is an assumption, educated guess, or conjecture as to what the completed research will reveal. The formulation of an hypothesis assumes some factual knowledge of the problem and research area. The more the researcher knows about the problem and the area, the more educated his or her guesses will be. However, if the research is concerned with a problem for which little or no information exists, the researcher may not be able to formulate any reasonable hypotheses or assumptions.

The hypothesis is a statement, often a question, that can be answered affirmatively, negatively, or "likely," after the research data have been collected and analyzed. Instead of being presented as a separate component of the overall research problem, the hypothesis may be implied in the problem statement. In this event, the objectives of the research may not be as precisely defined as desired. The hypothesis serves the purpose of clarifying and further defining the research problem. For example, if the problem is to determine the spatial distribution of soybean production and its relationships to the physical character of the land in area X, the hypothesis, if based on some factual information, might be stated, "Are soybeans cultivated exclusively on low slope ground where slopes do not exceed 5 degrees?" The completed research should confirm or deny this hypothesis, in total or in part.

If the problem, research area (matrix), and hypotheses are soundly constructed and clearly stated, the types of information needed will be self-evident. The researcher then must decide how to classify or categorize the data to be obtained. Classification systems must be based on meaningful criteria. Each class of data must be defined precisely and constructed so that observations and recording can be done in the field without confusion and undue difficulty. In many field problems, classification systems for specific phenomena have already been developed for the research area, or classification systems used in other similar areas may suffice as they are or in modified forms. The intellectual exercise of developing classification systems can be furthered by (1) a careful prefield study of existing systems and supplementary documents, (2) a reconnaissance of the research area, and (3) the development of broad classifications to be tested in the field in pilot areas and refined on the basis of the pilot studies. The classification systems, scale of the map base, and detail of the observations and recordings are closely interwoven. The *minimal areal unit*, which may be as small as a fraction of an acre in some problems or as large as several hundred acres in others, will determine the scale or level of resolution and detail of the data collected.

Assuming sound classification systems have been developed and the scale of the field work resolved, the actual collection of data can be undertaken. The acquisition of field data is accomplished in three major ways:

1. By mapping observable features and objects of the landscape;
2. By interviewing to obtain information concerning the nonvisible aspects of the area; and
3. By using instruments—stream gauges, climatological equipment, etc.—when precise information about a specific process is required.

All three methods can be used systematically and will provide accurate data that later may be analyzed quantitatively.

The nature of a specific problem, including the size of the research area, will determine the degree of coverage, or *intensity*, of the field work. Sometimes a given research area may be so large that total coverage by mapping or interviewing is not possible. Sampling procedures then must be employed. The use of instruments frequently implies that sample sites selected for observation as data could not be obtained for each square foot of the total area within a practical framework of time and cost. There is no standard formula that can be applied universally to all field problems to determine whether total coverage of an area is necessary or what types and amount of sampling should be utilized. The specific nature of the problem, types of data required, size of the research area, etc., as well as the degree of accuracy required, must be taken into consideration.

Primary or raw information must be compiled and organized for further analysis. Information on base maps must be measured, responses to interviews tabulated, and instrumentation data plotted or charted. If all the previous steps have been carefully planned and meticulously carried out, compilation and tabulation of the figures are normally simple tasks, although often time consuming. From this point on, further analysis of field data is not different from the methods of analysis used in other research problems. The methods of cartographic and quantitative and statistical analysis employed, and their degree of sophistication, depend upon the specific field problem undertaken. Geographic field methods and techniques focus on the systematic acquisition of field data and, although the completion of a field research problem requires detailed analysis, no special or unique methods of analysis are normally applied that are not used in other research problems.

The analysis of the field data, when completed, should answer, in total or in part, the questions raised in the statement of the problem as well as confirm or deny the hypothesis(es). As is true in other research, the answers to the questions may not be those that were anticipated, or the questions may remain unresolved. However, if the research design, methodologies, and techniques were sound, the results of the study will still be valuable. In research work it is not possible to answer the question or come to positive conclusions in all cases. In the final analysis, the success of a research problem is judged on the manner in which it was carried out rather than on whether the conclusions were positive, negative, or inconclusive.

## TRAINING IN FIELD GEOGRAPHY

The values and purposes of geographic field experience at the college level may be grouped into two broad sets of objectives:

1. To prepare the student for a career in applied geographic research.
2. To enrich the general and overall education of the student in geography and related fields.

### Preparation for a Career in Applied Research

Not all significant geographic research is based on primary data collected in the field. However, certain contemporary geographic research problems do depend, in part or in total, on information obtained by field study, and it is likely that this type of research will increase in the near future. It is becoming more and more evident that the research efforts of geographers can contribute significantly toward the solution of many kinds of pressing problems that face large numbers of areas in various parts of the world. Students who aspire to a career in this type of research must be knowledgeable and skilled in the methods and techniques of geographic field research. Equally important, they must be prepared to develop field research designs and to direct field work to be done by others. This includes a knowledge of what kinds of information can and need be obtained from field observations as well as how and at what scales. It also includes a clear understanding of the limitations and restrictions of field research in terms of time and money.

### Field Study and General Education

Field study has long been utilized as an instructional tool in the general education of geography students. An exposure to the real world to test basic geographic concepts learned in the classroom has been a widely used and accepted method of teaching at all levels. The values and objectives of field experience in geographic education are well stated by Professor John Fraser Hart, University of Minnesota:

1. To develop a better understanding of the nature of things discussed in the classroom and read about in books.
2. To enhance the student's ability to read the landscape, and to expose him both to the methods of the geographer and to such basic geographic concepts as spatial distribution, areal association, areal differentiation, spatial interaction, and the extremely difficult and complicated problem of generalization.
3. To enable the student to experience the thrill of personal discovery. Although many bright minds have been attracted into geography by recent emphasis upon its abstract and theoretical aspects, geographers should not lose sight of the discipline's traditional attraction for those to whom the empirical and concrete are congenitally more appealing.
4. To help the student learn to enjoy reading a landscape. Many people, including geographers, derive considerable intellectual stimulation and aesthetic satisfaction from working out of doors in contemplation of the complexity that is total reality. "Charles Darwin, after a most undistinguished undergraduate

career, was finally aroused by tramps through the countryside with his botany professor" (Anderson, 1959–1960, p. 8). A true field experience is a Socratic seminar in the open air, centering around observations, whether with or without the aid and guidance of an instructor.

5. To help the student learn to distinguish between necessary and extraneous information. This is the time in his undergraduate education when he is most likely to be called to think through the formulation of problems to do his own basic research, to collect and analyze data, and to put them into presentable form. Whether he wishes it or not, he is thrust into a personal learning situation, because he cannot depend upon others; in the field he must make his own decisions.*

## THE STRUCTURE OF FIELD COURSES

The basic emphases of contemporary field courses or courses that include field work may be contrasted in several ways:

1. Comprehensive versus specialized courses;
2. Courses designed for general education versus preparation for a professional career; and
3. Introductory versus advanced courses.

Comprehensive courses normally are completely devoted to field work and encompass methods, techniques, and subject matter pertinent to several major subfields of the discipline, both cultural and physical. Frequently they consist of exercises or problems concerned with a variety of topics, such as landforms, natural vegetation, agriculture, urbanization, manufacturing, recreation, and perception. Generally they do require a general background in geography, but not previous field training. Therefore, they are offered at the advanced undergraduate or graduate levels. Specialized field courses emphasize techniques and content applicable only to one or two subfields. Often the field work is only a part of a total course such as climatology, advanced urban geography, etc., and the field work usually focuses on problems in microclimatology, or urbanization processes, or population mobility, etc. These courses are designed for advanced undergraduate or graduate students who have acquired a degree of specialization in the proper area of study.

A course designed for general education may cover field work exclusively or have field study as only a part. Such courses usually are offered at the lower undergraduate level and require no previous field training and little background in geography. In some instances, they are offered at the freshman level. General rather than sophisticated field techniques are stressed, and the primary objectives are to supplement concepts and ideas acquired elsewhere and to provide opportunities to learn the basic skills of map reading, identification of features, and so forth. Courses designed to prepare students for professional careers are normally graduate level courses and emphasize sophisticated methods and techniques. Usually previous field experience is required as well as a substantive background in

---

*"The Undergraduate Field Course," *Field Training in Geography,* Technical Paper No. 1, Commission on College Geography, Association of American Geographers, 1968, p. 29.

geographic methodology. The development of the latter type of course is difficult in that it requires a great deal of planning on the part of the faculty.

Introductory courses require little or no background in geography or field work. They are offered as a separate course or as a part of a course and usually are semi-comprehensive. Advanced courses imply previous field training and a degree of specialization. The course emphases described are not mutually exclusive and differ in degree rather than type. Introductory courses as well as certain advanced courses may be general-education oriented and comprehensive in scope. There are numerous combinations, and the most effective course design depends largely on the objectives of the institution, the size and goals of the department, the interests and backgrounds of the faculty, and the type and career goals of the student body.

## Field Course Formats

Field courses are organized in several different ways. There are courses that are completely devoted to field study, and there are courses in which field study is only a part of the course. The former has two generally accepted formats: the off-campus (field-camp) course and the on-campus course. The off-campus course is a five- to eight-week period, usually during the summer, devoted exclusively to field study and free from campus commitments and distractions. It provides an unusual academic experience in that students and faculty are in continuous close contact for a pro-longed period to concentrate their efforts on the specific tasks at hand. Further, working in an environment different from the base campus provides the students with the opportunity to make comparisons between the two areas. The major drawbacks are costs in time and money. Problems of housing, food, travel, schedul-ing, etc., must be resolved. Because the on-campus course is scheduled during the academic year, demands of other courses and campus activities are a serious disadvantage. In addition, continuity is difficult to maintain, and field exercises that require a continuous prolonged effort must be eliminated.

With careful planning it may be possible to combine off- and on-campus activities to maximize the advantages and to minimize the disadvantages of both. For instance, certain field problems might be scheduled for a week or ten days during a semester or quarter break. The remainder of the course would be con-ducted on campus. Regardless of the specific format, whether off-campus, on-campus, or a combination, situations that reduce time constraints allow for a variety and diversity of projects.

Field work as part of a course is common in many advanced courses. However, the field work is focused on specialized problems compatible with the content and objectives of the overall course. Field study is highly compatible with a large number of advanced courses, particularly those concerned with research methodologies. Field problems can be included without impairing the objectives and structure of the course and, in most cases, improve its relevancy and practicability. Such field work usually takes place near or on the campus and may consist of only one or two field exercises.

The inclusion of field work in introductory courses is also quite common. In these cases, the field work is general and elementary rather than specialized, since most of the students have no previous relevant experience. The large number of students usually enrolled in elementary courses further restricts the kind of field work that can be accomplished.

**The Field Trip and the Field Excursion.** The field trip and the field excursion are widely used methods of teaching. The field trip normally refers to a study of a few hours to several days, while the field excursion usually requires several weeks as well as extensive travel. The emphasis of either can be very specific or as comprehensive as time and materials will allow. Field trips and excursions may be associated with a specific course or may be open to all interested persons and offered with or without credit. Their primary function is to serve as an instructional device rather than as a research tool. The major objectives of a field trip or excursion are:

1. To demonstrate basic geographic concepts learned in the classroom,
2. To provide an environment for the student to acquire new knowledge and perceptions,
3. To stimulate the student's curiosity and imagination, and
4. To serve as a mechanism to develop an orderly and systematic manner of observation and to relate these observations to an integrated whole.

A carefully planned and well-organized field trip or excursion can be a highly valuable learning experience.

**The Graduate Seminar.** The graduate seminar is the most advanced type of field course. It requires previous formal field training and assumes that the participants have the skills and knowledge of the basic methodologies and techniques of data collection. Further, it implies a sound background in one or more areas of specialization. The primary purpose of the graduate seminar is to prepare students for professional work or advanced graduate training. Emphasis is placed on research design, sophisticated data gathering techniques, and critique. In many cases, team research is stressed. The most effective graduate seminars are two to three weeks in length and are conducted off campus to provide a continuous and intensive learning experience. The graduate seminar is normally the final formalized field instruction that the student will receive before undertaking professional work or advanced graduate research.

### Cost and Requirements of Field Instruction

Good field instruction is costly in terms of dollars and faculty time. Supplementary materials such as base maps, field instruments, and supplies must be obtained, and travel and insurance expenses will be incurred. However, field training is not unique to geography. Field work is an integral part of the training of anthropologists, botanists, geologists, sociologists, and zoologists, as well as other professional fields. The costs of field training in geography, however, are not appreciably greater than those of essential laboratory work in chemistry, physics, or biology; therefore, if the field serves as the laboratory for certain types of geographic study, a rationale for expenditure has been established.

Perhaps the greatest cost of field instruction is faculty time. Field exercises must be developed, potential field sites selected and checked out, trips organized, and accommodations reserved. These are but a few of the additional efforts required for field courses. In addition, the number of student-instructor contact hours is usually greater than in equivalent credit nonfield courses. Also, student problems are more numerous. Planning and preparation play a more critical role in field instruction than in other types of instruction.

## THE COMPREHENSIVE FIELD COURSE

Regardless of the emphasis or format employed, effective training in field geography must be problem oriented. There must be a clearly defined purpose and stated goals for the field course as a whole as well as for each specific field exercise or problem.

The ideal comprehensive field course consists of a variety of field exercises and problems. The rationale is that the student exposed to a diversity of field methods and techniques develops skills and perspectives that are applicable to a wide range of field situations. From this acquired repertoire of techniques, the student can select, refine, and apply approaches to meet the needs of most given problems. One field course will not make a student a master of all phases of data gathering, but he or she will be knowledgeable in the sense of knowing how and when to employ a given technique.

Field exercises should be designed to represent various stages of difficulty and complexity. The only objective of a given exercise might be to acquire skills in the techniques of data collecting; another exercise might include both data gathering and analysis, with emphasis on the analysis; while other exercises might stress the development of the total research design, including decisions about what data are essential, what techniques should be used to acquire the data, and what methods of analysis should be employed.

Initially, highly structured simple data-collection exercises should be given. Subsequent exercises should become more complex and open-ended, until the student finally creates the entire research framework.

Some exercises are best done in groups of two or three, with students learning from each other, whereas other problems are better suited for one individual. Certain exercises could involve the entire class, with each student completing a component of the total problem (see Appendix A).

Dividing each field exercise or problem into three distinct phases has proved effective. The first and third phases involve the entire class and may be done in the classroom. The first phase sets the stage for the forthcoming field work. At this time, the exercise is discussed and put into its proper setting. For example, if the exercise is concerned with agricultural land use and agricultural economics, the class discusses the history, scope, and methodologies of agricultural geography as well as how this particular exercise fits into the broad perspective. This prefield briefing would include instructions as to how the exercise should be done. The instructions might be highly detailed or very brief depending upon the objectives of the particular exercise.

The second phase consists of the actual field work done by teams of students or by individuals. This includes the preparation of some type of tangible end product in the form of maps and tables or a written report.

The third phase consists of an analysis of the reports. At this time, problems encountered, possible new approaches to the exercise or similar exercises, and a general assessment and critique of the work completed may be discussed.

This suggested structure for the comprehensive field course has proved to be successful, but it is only one of several approaches. It may be modified with ease to meet the needs of a given area and institution as well as to maximize the background and interests of the particular instructor(s). There is no (nor should there be) "standard" comprehensive field course. Each area, institution, class, and instructor is unique, and the effective field course should be adapted accordingly.

## SUGGESTED REFERENCES

### NATURE AND SCOPE OF GEOGRAPHIC FIELD RESEARCH

ACKERMAN, E. A., *Geography as a Fundamental Research Discipline,* Department of Geography Research Paper No. 53. Chicago: University of Chicago Press, 1958.

BROWN, E. H., "The Teaching of Fieldwork and the Integration of Physical Geography," *Trends in Geography: An Introduction Survey,* edited by R. V. Cooke and J. H. Johnson, 70–78. London: Pergamon Press, 1969.

DAUGHERTY, RICHARD, *Science in Geography: Data Collection.* London: Oxford University Press, 1974.

DAVIS, CHARLES M., "Field Techniques," *American Geography: Inventory and Prospect,* edited by P. E. James and C. F. Jones, 496–529. Syracuse, New York: Syracuse University Press, 1954.

FRIBERG, J. C., *Field Techniques and the Training of the American Geographer,* Discussion Paper No. 5, Department of Geography, Syracuse University, 1975.

HARING, L. L. and J. F. LOUNSBURY, *Introduction to Scientific Geographic Research.* Dubuque, Iowa: William C. Brown Co., Publishers, 1983.

HART, J. F., ed., *Field Training in Geography,* Commission on College Geography Technical Paper No. 1. Washington, D.C.: Association of American Geographers, 1968.

PLATT, ROBERT S., *Field Study in American Geography: The Development of Theory and Method Exemplified by Selections,* Department of Geography Research Paper No. 61. Chicago: University of Chicago Press, 1959.

SAUER, C. O., "The Education of a Geographer," *Annals of the Association of American Geographers,* September 1956, 287–99.

STODDARD, R. H., *Field Techniques and Research Methods in Geography.* Dubuque, Iowa: Kendall/Hunt Publishing Co., 1982.

*THE SCIENCE OF GEOGRAPHY,* Report of the Ad Hoc Committee on Geography, Earth Science Division, National Academy of Sciences-National Research Council, Publication 1277. Washington, D.C.: National Academy of Science-National Research Council, 1965.

HISTORICAL DEVELOPMENT: CASE STUDIES

HUDSON, G. D., "The Unit Area Method of Land Classification," *Annals of the Association of American Geographers,* June 1936, 99–112.
JONES, C. F. and R. PICO, eds. *Symposium on the Geography of Puerto Rico.* Rio Piedras, Puerto Rico: University of Puerto Rico Press, 1955.
*RURAL LAND CLASSIFICATION PROGRAM OF PUERTO RICO,* Northwestern University Studies in Geography, No. 1, 1952.
STAMP, L. D., *The Land of Britain: Its Use and Misuse.* London: Longmans Group Ltd., 1948.

# Base Maps for Field Studies

## Chapter

# 2

**Chapter**

**2**

A major part of geographic field work is concerned with the spatial recording of landscape features—physical, cultural, and economic—that can be observed and classified or categorized. A well-organized field mapping problem consists of five distinct phases, which must be completed in order. These phases are as follows:

1. Determination of precisely what data are essential to the solution of the specific field problem.
2. Selection of the scale or level of resolution of the data to be obtained.
3. Choice of the classification system(s) to be employed in categorizing the data.
4. Selection of most effective base map for the specific problem.
5. Actual field mapping.

## LINEAR AND AREAL MEASUREMENT

The United States is in the process of converting from the English to the metric system. At the time of this writing, however, many of the tools used in field work (base maps such as topographic quadrangles and plat maps, legal descriptions of property, etc.) are still based on the English system. The basic units of measurement used in practically all field investigations are concerned with linear and areal calculations. The public land surveys, dating back to 1785, established the township-range system which covers most of the country and serves as the basis of land/property ownership. Due to the historical and rigid legal precedents, conversions to the metric system likely will not take place for many years. For this reason, the English system, as it pertains to linear and areal measurement, will be used in this text. However, in areas outside of the United States, the metric system is, or soon will be, the basic unit of measurement. Therefore, the field researcher must have a working knowledge of both systems and be prepared to use whichever one is relevant to the specific area of research and time (see Tables 2-1 and 2-2).

## BASE MAPS

Mapping implies recording facts, drawing boundaries, and making notes on something. This "something," if it contains an adequate number of orientation or reference points that spatial or areal extent of field observations can be recorded accurately, is a *base map*.

The selection of the base map upon which the data will be recorded in the field is highly critical. The base map determines to some extent how the field data are to be analyzed and presented in their final form, both statistically and

### TABLE 2-1.

*Linear Measure.*

| Inches/Centimeters | | Yards/Meters | | Miles/Kilometers | |
|---|---|---|---|---|---|
| 1 | 2.54 | 1 | 0.914 | 5 | 8.047 |
| 2 | 5.08 | 2 | 1.82 | 10 | 16.09 |
| 3 | 7.62 | 3 | 2.74 | 15 | 24.14 |
| 4 | 10.16 | 4 | 3.65 | 20 | 32.18 |
| 5 | 12.70 | 5 | 4.57 | 25 | 40.23 |
| 6 | 16.24 | 6 | 5.48 | 30 | 48.27 |
| 7 | 17.78 | 7 | 6.39 | 35 | 56.32 |
| 8 | 20.32 | 8 | 7.31 | 40 | 64.36 |
| 9 | 22.86 | 9 | 8.22 | 45 | 72.42 |
| 10 | 25.40 | 10 | 9.14 | 50 | 80.45 |
| 11 | 27.94 | 11 | 10.05 | 55 | 88.51 |
| 12 | 30.48 | 12 | 10.97 | 60 | 96.54 |
| 13 | 33.02 | 13 | 11.88 | 65 | 104.61 |
| 14 | 35.56 | 14 | 12.80 | 70 | 112.63 |
| 15 | 38.10 | 15 | 13.71 | 75 | 120.70 |
| 16 | 40.64 | 16 | 14.62 | 80 | 128.72 |
| 17 | 43.18 | 17 | 15.54 | 85 | 136.79 |
| 18 | 45.72 | 18 | 16.45 | 90 | 144.81 |
| 1 in. = 2.54 cm | | 1 yd = 0.914 m | | 1 mi = 1.6093 km | |
| 1 cm = 0.3937 in. | | 1 m = 1.093 yd | | 1 km = 0.62137 mi | |

TABLE 2-2.

*Area (Square) Measure.*

| Square Feet | / | Square Meters | Acres/Hectares | | Square Miles | / | Square Kilometers |
|---|---|---|---|---|---|---|---|
| 1 | | 0.0929 | 1 | 0.40469 | 1 | | 2.590 |
| 2 | | 0.1858 | 2 | 0.80937 | 2 | | 5.180 |
| 3 | | 0.2787 | 3 | 1.2141 | 3 | | 7.770 |
| 4 | | 0.3716 | 4 | 1.6188 | 4 | | 10.36 |
| 5 | | 0.4645 | 5 | 2.0234 | 5 | | 12.95 |
| 6 | | 0.5574 | 6 | 2.4281 | 6 | | 15.54 |
| 7 | | 0.6503 | 7 | 2.8328 | 7 | | 18.13 |
| 8 | | 0.7432 | 8 | 3.2375 | 8 | | 20.72 |
| 9 | | 0.8361 | 9 | 3.6422 | 9 | | 23.31 |
| 10 | | 0.9290 | 10 | 4.0469 | 10 | | 25.90 |
| 11 | | 1.022 | 11 | 4.4516 | 11 | | 28.49 |
| 12 | | 1.115 | 12 | 4.8562 | 12 | | 31.08 |
| 13 | | 1.208 | 13 | 5.2609 | 13 | | 33.67 |
| 14 | | 1.301 | 14 | 5.6656 | 14 | | 36.26 |
| 15 | | 1.394 | 15 | 6.0703 | 15 | | 38.85 |
| 16 | | 1.486 | 16 | 6.4750 | 16 | | 41.44 |
| 17 | | 1.579 | 17 | 6.8797 | 17 | | 44.03 |
| 18 | | 1.672 | 18 | 7.2844 | 18 | | 46.62 |

1 ft$^2$ = 0.0929 m$^2$       1 acre = 0.404687 hectare       1 m$^2$ = 2.590 km$^2$

1 yd$^2$ = 0.8361 m$^2$       1 hectare = 2.4710 acres       1km$^2$ = 0.3861 m$^2$

1 m$^2$ = 10.764 ft$^2$       1 hectare = 10,000 m$^2$       1 km$^2$ = 247.104 acres

cartographically. To select the most desirable base map for a given problem, the following should be considered:

1. The nature and detail of the data that are to be collected.
2. The scale "minimum areal unit" at which the data are to be mapped.
3. The types of mapping techniques to be employed: single-feature or multiple-feature, total mapping coverage or transect or sample-area mapping, etc.
4. The system(s) of data classifications to be used.
5. The types and number of control or reference points that a given type of base map will provide.

There are several types of maps or maplike material that serve as adequate base maps. The most common and readily available are:

1. The conventional vertical aerial photograph,
2. Topographic maps (quadrangles), and
3. Plat (cadastral) maps.

Each of these "standardized" base maps has advantages and limitations. The specific nature of a given field problem determines which should be used. Other materials may serve as adequate base maps, including some types of remotely sensed imagery other than aerial photographs. When no suitable base maps exist—usually when the field problem is microscale in nature—base maps may have to be constructed. In this event, plane table methods are normally employed.

## Vertical Aerial Photographs

Conventional *vertical aerial photographs* often are the most desirable base maps for many kinds of field problems at the intermediate or microscale. They are widely used for urban and rural land-use studies as well as in physical and cultural field investigations. A vertical aerial photograph is a photograph taken with the camera pointing vertically downward. Oblique aerial photographs are photographs taken with the optical axis of the camera noticeably inclined from the vertical. Oblique photography does not show standard scale over the entire photo (scale is constant only along any one line parallel to the horizon, but decreases in geometric progression from foreground to background) and therefore cannot be used for precise field mapping.

The conventional vertical aerial photograph is not a map but rather an actual picture of reality. As a result, a large number and variety of control or orientation points are provided—surface landforms (drainage patterns, hills, knolls, valleys, gullies, etc.), vegetation (woodlots, tree lines, cultivated crops, etc.), artificial structures (roads, lanes, dwellings, farm buildings, drainage ditches, etc.), and soil types illustrated by differences in shading or color (see Figures 2-1 and 2-2).

Vertical aerial photographs are taken with the airplane flying a course as level as possible at a given elevation and with the optical axis of the camera perpendicular to the ground. The point on the ground shown on the photo that is precisely perpendicular to the camera is known as the *principal point,* or *center point.* It is normally indicated on the photograph as a small cross (see Figure 2-3). The minimum of distortion occurs in the central portion of the photograph surrounding the principal point. The marginal areas of the photograph are distorted in scale, direction, and relief due to parallactic displacement, or *parallax.* The outer margins of the photograph are not suitable for accurate mapping. In general, the central 50 percent of an individual photograph is sufficiently free of parallax distortion for precise mapping and measurement.

Aerial photographs are normally taken in parallel north-south overlapping strips (about 60 percent overlap in the area between adjacent photos in a given flight strip and about 30 percent sidelap in the area between photos of adjoining flight strips). Such coverage allows the field worker to plot the usable area on each photo prior to the actual field mapping and avoid using the distorted marginal areas.

For certain field problems, flight lines may be planned to connect specific points that are not true north-south oriented. Also, specific field research work may need a higher percentage of overlap than is conventional, and flight lines can be planned accordingly.

Vertical aerial photos are available for each county in the United States. Several governmental agencies and private commercial concerns maintain libraries and serve as distributors. The Agricultural Stabilization and Conservation Service

**Figure 2-1.**

Portion of an Enlarged Conventional Vertical Aerial Photograph (*scale: 1 inch represents 2000 feet*).

Note the large number and variety of control or reference points—field patterns, roads and lanes, dwellings and farmsteads, land use, tree lines, and so forth. Many of these control points are lacking on topographic and other similar maps. At this scale, the minimum mapping unit would be about 200 linear feet. (Courtesy U.S. Department of Agriculture)

## Figure 2-2.

Portion of an Enlarged Conventional Vertical Aerial Photograph
*(scale: 1 inch represents 800 feet).*

This photograph covers the portion of the same
area as that illustrated in Figure 2-1. At this
scale, the minimum mapping unit would be about
100 linear feet. Controlled mapping is increased
as additional control or reference points are provided. (Courtesy U.S. Department of Agriculture)

---

*It is common to use imagery at the scale of one inch = 400 feet or larger for field work.

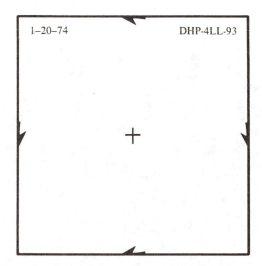

## Figure 2-3.

Each aerial photograph has collimating or fiducial marks in the form of various types of marks at the center of each side of the photo. Normally these marks are shown as arrows. In addition, the principal point, or center point, indicated by a small cross, is the intersection between the marginal arrows that are often indicated. It is the optical center of the photo or the point directly perpendicular to the camera lens.

Each photo is marked by numbers indicating the code or symbol number, roll number, and exposure number of the negative. The date the photo was taken is also indicated.

| | |
|---|---|
| DHP | Code or symbol of the overall area of coverage |
| 4LL | Roll number that identifies flight strip |
| 93 | Exposure number of specific photo |
| 1-20-74 | Date of photo (January 20, 1974) |

(ASCS), U.S. Department of Agriculture, has coverage available at various scales for all counties in the country. Several standardized scales that normally are available are shown in Table 2-3.

There is a photo index sheet(s) for each county in the United States which gives the code numbers and dates of the individual photos. The index sheet must be referred to in order to determine what photos cover what specific areas (see Figure 2-4).

Conventional aerial photographs are extremely useful because they provide a large number and variety of control or reference points. Using these photos, the field worker can determine without difficulty where he or she is at any given time and can accurately plot whatever data are being mapped. However, a vertical photo can be considered a true representation of a segment of the earth's surface only under ideal conditions. These conditions are absolute flat terrain, precise level flight course, exact vertical positioning of the optical axis of the camera,

TABLE 2-3.

*Conventional Scales of Aerial Photographs.*

| Type of Reproduction | Paper Size | Approx. Scale from 1:20,000 Photography | Approx. Scale from 1:40,000 Photography |
|---|---|---|---|
| Photo index* | 20" × 24" | | |
| Contact print** | 9½" × 9½" | 1" = 1667' | 1" = 3334' |
| Enlargement | 12" × 12" | = 1320' | = 2640' |
| Enlargement | 12" × 12" | | = 1320' (sectional) |
| Enlargement | 17" × 17" | = 1000' | = 2000' |
| Enlargement | 17" × 17" | | = 1000' (sectional) |
| Enlargement | 24" × 24" | = 660' | = 1320' |
| Enlargement | 24" × 24" | = 330' (sectional) | = 660' (sectional) |
| Enlargement | 38" × 38" | = 400' | = 800' |
| Enlargement | 38" × 38" | = 200' (sectional) | = 400' (sectional) |

*Number of photo index sheets per county depends on the size of the county.
**Contact prints are not available as sectionals or with scale accuracy.

and absence of differential shrinking of film or paper. In addition to the parallax distortion along the margins of the photo previously discussed, possible distortion might result from other factors, such as:

1. Irregular topography, which places hills and highlands closer to the camera and valleys and lowlands a greater distance away from the camera lens. The photograph scale is the ratio of camera focal length to camera height ($f/H$). Since the focal length is consistent in a given camera, changes in the height of the camera above the ground will result in scale changes. In addition, high points will be displaced outward from the center of the photograph. Conversely, below a selected base plane, low points will be displaced radially inward. Technically, topography-caused image displacements can be removed to a large degree by producing an *orthophotograph*, which is essentially a transformation of a photograph into a planimetric image.
2. Variations in flight altitude, which affect scale from one photo to another. These differences in scale may be corrected by properly enlarging or reducing the photo during the printing process.
3. Significant tilting of the optical axis of the camera which results in an oblique image rather than a true vertical photo. Tilting in excess of 4 degrees will result in errors that normally prohibit the use of the photo for precise mapping and measurement.
4. Differential shrinkage of photographic paper, which could be sufficient to interfere with precise measurements. However, special low-shrinkage paper may be used where a high degree of scale accuracy is desired.
5. Lens distortions (chromatic, spherical, radial, astigmatic, etc.), which are expected in ordinary hand cameras but are normally negligible in cameras designed specifically for aerial photographic work.

## Figure 2-4.

Portion of a Photo Index Sheet for Maricopa County, Arizona (*scale: 1 inch represents 2 miles*).

The number of photo index sheets for a given county depends upon the areal size of the county. Figure 2-4 represents about one-fourth of one photo index sheet, and there are fifteen photo index sheets that cover Maricopa County. Each county ASCS Office has similar photo indexes which give the code numbers and date of flight for individual photos in the county. These code numbers, as well as an order blank, must be obtained before photographs can be purchased. (Courtesy U.S. Department of Agriculture)

A major advantage of vertical aerial photos is that certain types of phenomena, after adequate field checking has confirmed differences in color and shading, may be mapped in the office or laboratory. To be effective, however, the photos must be recent, and the type and classification of data desired of a scale easily differentiated by shadings and color.

Often, laboratory work is facilitated by the use of a *stereoscope*. Adjoining photos, each covering some common ground area, can be viewed through a stereoscope, either the lens (refraction) or mirror (reflection) type, which will provide a three-dimensional picture of the area viewed. Differences in topography are highly exaggerated, and the extent of relief is far greater than that observed visually from an airplane. The conventional flight-strip photographs are normally taken hundreds (sometimes thousands) of feet apart along the flight line. Therefore, the much greater angular difference between lines from two successive exposure locations to a common ground point (as compared with the lines from an aerial observer's eyes to the same ground point) account for the exaggerated vertical relief when viewed stereoscopically (see Figure 2-5). Stereoscopic analysis is useful both in the laboratory and in the field for acquiring an overall view of the field-mapping area. In addition, stereoscopic vision may highlight slight differences of texture, tone, pattern, etc., on the photographs, which may be useful in the precise determination of vegetative and soil types and boundaries. In certain cases, stereoscopic analysis may be used for precision field mapping and measurement.

On occasion, the wealth of information provided by a vertical aerial photo may be a disadvantage in certain field mapping problems, particularly if single-feature mapping is utilized. The aerial photo is not selective, and all visible features of the landscape are shown. If a field problem is concerned with mapping slope only, for example, the vegetation, cultural features, etc., that appear are superfluous and may interfere with the slope observations and recording.

## Topographic Maps (Quadrangles)

*Topographic maps*, or *quadrangles*, differ from aerial photographs in that not all visible features of the landscape are illustrated. Topography, drainage patterns, and major artificial features appear, but not land use, minor topographic features (relief so small that it falls between contour lines), detailed vegetative cover, and soil types. Also included are such nonvisible phenomena as land-survey and political boundaries, precise elevation points or lines, and identification of types of natural and artificial features (see Figure 2-6).

The U.S. Geological Survey has been producing topographic maps for almost a century, and much of the continental United States, Alaska, Hawaii, and Puerto Rico have been mapped. Individual maps are published as part of a map series. A map series is a group of maps produced according to the same overall specifications and usually having a common scale and symbol system. The Topographic Series is comprised of several series (sometimes referred to as sets) of maps. The dimensions of each map within a series are established by standard latitudinal and longitudinal angles between selected meridians and parallels. A common symbolism is also employed. The principal series comprising the Topographic Series are listed in Table 2-4. In addition to these principal series, a group

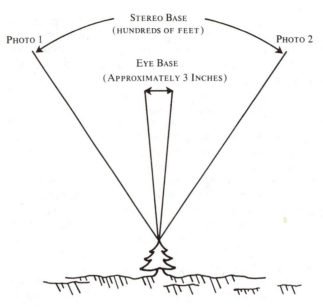

**Figure 2-5.**

The eye base (distance between the pupils of the eyes) is normally about 3 inches (7.5 centimeters). The stereo base (distance between common image points on adjacent photos) may be several hundred feet. The vertical scale in viewing a stereo pair of photographs is greatly exaggerated due to the much greater angular difference between lines of sight from two widely spaced photographs to a given ground point as compared to the line of sight from an observer's eyes to the same ground point.

covering national parks, monuments, and historical sites is available, as well as a set for certain metropolitan areas, a set for rivers and floodplains, and a set for Puerto Rico and other areas.

A complex system of symbols depicts natural and artificial features. Colors are an integral part of the symbolization. Black is used for artificial features; blue for water features; brown for relief; green for vegetation; purple on interim maps for areas that have been revised since the earlier edition; and red for urban areas, some roads, and land boundaries (see Table 2-5).

Topographic maps show the township and section land boundaries based on the public land surveys dating back to 1785. The public land surveys were based on thirty-one pairs of *base lines* (parallels connecting points of equal latitude) and *principal meridians* (lines connecting points of equal longitude) for the continental United States and on five similar pairs for Alaska. Base lines served as reference lines for rows of townships running east-west, each 6 miles in width. The principal meridians served as reference lines for ranges running north-south, each approximately 6 miles in width. Each township and range is numbered and identified. The

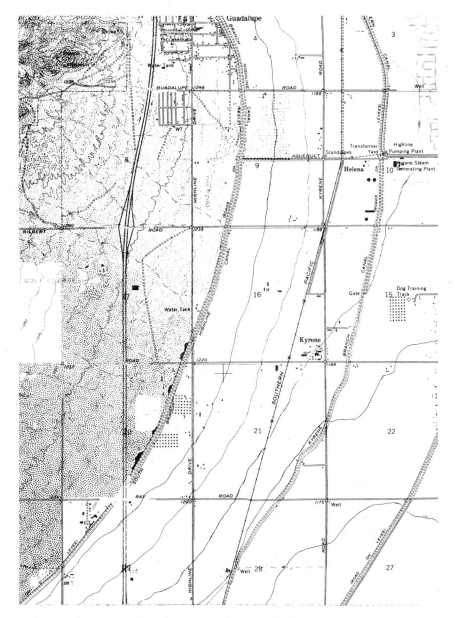

## Figure 2-6.

Reduced Portion of a Topographic Quadrangle [7.5-minute series (topographic), Guadalupe, Arizona].

The map covers the areas shown in Figures 2-1 and 2-2. Note the absence of minor ground features found on the vertical aerial photographs. However, section line boundaries and the identification of major cultural and natural features, lacking on the aerial photographs, may be used as control points in field mapping. At the original (unreduced) scale of this map, 1:24,000 (1 inch = 2000 feet), the minimum mapping would be about 200 linear feet. (Courtesy U.S. Geologic Survey)

TABLE 2-4.

*Principal Map Series (Sets) in the Topographic Map Series, U.S. Geological Survey.*

| Series/Sets | Scale | One Inch Represents (in miles) | Angular Size (latitude-longitude) (in minutes) | Area Covered (depends on latitude) (in miles) |
|---|---|---|---|---|
| 7½-minute | 1:24,000* | 0.379 | 7½ × 7½ | 49 to 71 |
| 15-minute | 1:62,500 | 0.986 | 15 × 15 | 197 to 292 |
| Alaska | 1:63,360 | 1.0 | 15 × 20 to 30 | 207 to 281 |
| United States | 1:250,000** | 3.95 | 1° × 2° | 4580 to 8669 |
| State Maps | 1:500,000 | 7.89 | State coverage | State coverage |
| United States | 1:1,000,000 | 15.78 | 4° × 6°*** | 73,734 to 102,759 |

*For Alaska, the scale is 1:25,000, and for Puerto Rico, 1:20,000.
**Maps of Alaska and Hawaii vary from these standards.
***For Alaska, the size is 4° × 12°.

*township*, approximately 36 square miles, serves as the basic land unit. Due to the curvature of the earth, the meridians converge at the poles; therefore, townships narrow as they go northward. To prevent excessive narrowing, *standard parallels*, usually placed four townships apart (24 miles), are used for reference lines. This results in an offset, or *jog,* at each standard parallel to maintain the six-mile-square size (see Figure 2-7). Each township is divided into 36 *sections*, each about one square mile in area, which in turn may be subdivided into *quarter sections*; the quarter sections may be subdivided into *quarter-quarters.* Each section also is subdivided into 16 parts called *legal subdivisions.* Each legal subdivision of 40 acres is the basic unit for development purposes (see Figure 2-8).

The township and range system is the basis of legal descriptions of property over much of the United States. Field researchers concerned with any aspect of land use or land development must be familiar with the system in order to collect and analyze field data effectively.

There were nineteen states (including the original thirteen) that were not included in the public land system. These states, except Texas, were all in the East or Southwest. The eastern states used a system of land survey known as *metes and bounds.* Each parcel of land was defined by means of natural (and/or artificial) marks such as ridges, streams, margins of wooded areas, and so forth. The metes-and-bounds survey was also used in areas of the Midwest that were set aside as military reserves prior to the public land survey. Much of the southwestern part of the United States, formerly under the control of Spain, was subdivided on the basis of a Spanish unit of length called the *vara.* Varas of three slightly different lengths were used in the southwestern areas: one vara in Texas = 33.34 inches, one vara in California = 33 inches, and one vara in Mexico = 32.99 inches.

The margins of topographic maps contain a great deal of information that is helpful in the field. This information includes the following: name of the map sheet and map series of which the map sheet is a part; location of the area covered, in

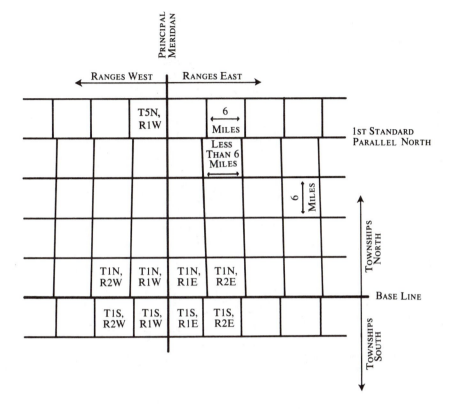

**Figure 2-7.**

The location of each township in a survey region
is based on the intersection of the base line and
principal meridian. Townships are numbered in order north and south of the base line, and the
number of the township east or west of the principal meridian is called the range. Jogs along
range lines at the standard parallels and base lines
offset the convergence of the meridians.

terms of latitude and longitude and location of the map in the state; date of edition;
date on revision; scale, contour interval; compass declination; type of projection;
notes on symbolization; reference to other grid systems; statement of map accuracy;
and names of adjoining map sheets.

The use of topographic maps for field mapping depends largely upon
the scale being employed for data recording and the kinds and number of control or
reference points a given map provides. The 7½-minutes series contains the most
detailed maps readily available (1 inch = 2000 feet). The minimum mapping area
would approximate 200 linear feet and as such would not suffice for microlevel field
work. Depending on the area covered, easily identifiable control points in the field
will vary considerably from one map to another. Flat terrain with few or no artificial
features probably would not provide sufficient control for accurate field mapping. If

## TABLE 2-5.

*Topographic Map Symbols, U.S. Geological Survey Topographic Maps.*

| | |
|---|---|
| Primary highway, hard surface | Boundary: national |
| Secondary highway, hard surface | State |
| Light-duty road, hard or improved surface | county, parish, municipio |
| Unimproved road | civil township, precinct, town, barrio |
| Trail | incorporated city, village, town, hamlet |
| Railroad: single track | reservation, national or state |
| Railroad: multiple track | small park, cemetery, airport, etc. |
| Bridge | land grant |
| Drawbridge | Township or range line, U.S. land survey |
| Tunnel | Section line, U.S. land survey |
| Footbridge | Township line, not U.S. land survey |
| Overpass—Underpass | Section line, not U.S. land survey |
| Power transmission line with located tower | Fence line or field line |
| Landmark line (labeled as to type) *TELEPHONE* | Section corner: found—indicated |
| | Boundary monument: land grant—other |
| Dam with lock | |
| Canal with lock | Index contour    Intermediate contour |
| Large dam | Supplementary cont.    Depression contours |
| Small dam: masonry — earth | Cut — Fill    Levee |
| Buildings (dwelling, place of employment, etc.) | Mine dump    Large wash |
| School—Church—Cemeteries    Cem | Dune area    Tailings pond |
| Buildings (barn, warehouse, etc.) | Sand area    Distorted surface |
| Tanks; oil, water, etc. (labeled only if water)    Water Tank | Tailings    Gravel beach |
| Wells other than water (labeled as to type)    Oil    Gas | |
| U.S. mineral or location monument — Prospect | Glacier    Intermittent streams |
| Quarry — Gravel pit | Perennial streams    Aqueduct tunnel |
| Mine shaft—Tunnel or cave entrance | Water well—Spring    Falls |
| Campsite — Picnic area | Rapids    Intermittent lake |
| Located or landmark object—Windmill | Channel    Small wash |
| Exposed wreck | Sounding—Depth curve *10*    Marsh (swamp) |
| Rock or coral reef | Dry lake bed    Land subject to controlled inundation |
| Foreshore flat | |
| Rock: bare or awash | |
| | Woodland    Mangrove |
| Horizontal control station | Submerged marsh    Scrub |
| Vertical control station    BM ×671 ×672 | Orchard    Wooded marsh |
| Road fork — Section corner with elevation    *429*  +*58* | Vineyard    Bldg. omission area |
| Checked spot elevation    × *5970* | |
| Unchecked spot elevation    × *5970* | |

Variations will be found on older maps.

Map symbols shown on U.S. Geological Survey topographic maps are in color. The color code is:

| | |
|---|---|
| Black | Artificial and cultural objects and features (roads, railroads, buildings, boundaries, quarries, cemeteries, airports, etc.) |
| Blue | Water features (streams, lakes, springs, wells, glaciers, marshes, and canals) |
| Brown | Topography (relief features and contour lines) |
| Green | Vegetation |
| Purple | On interim maps, areas that have been revised since the previous edition |
| Red | Urban areas, land boundaries, and some roads |

**38**

## UNITED STATES

| 6 | 5 | 4 | 3 | 2 | 1 |
|---|---|---|---|---|---|
| 7 | 8 | 9 | 10 | 11 | 12 |
| 18 | 17 | 16 | 15 | 14 | 13 |
| 19 | 20 | 21 | 22 | 23 | 24 |
| 30 | 29 | 28 | 27 | 26 | 25 |
| 31 | 32 | 33 | 34 | 35 | 36 |

## CANADA

| 31 | 32 | 33 | 34 | 35 | 36 |
|---|---|---|---|---|---|
| 30 | 29 | 28 | 27 | 26 | 25 |
| 19 | 20 | 21 | 22 | 23 | 24 |
| 18 | 17 | 16 | 15 | 14 | 13 |
| 7 | 8 | 9 | 10 | 11 | 12 |
| 6 | 5 | 4 | 3 | 2 | 1 |

A. CONVENTIONAL SYSTEMS OF NUMBERING SECTIONS IN THE UNITED STATES AND CANADA.

| 13 | 14 | 15 | 16 |
|---|---|---|---|
| 12 | 11 | 10 | 9 |
| 5 | 6 | 7 | 8 |
| 4 | 3 | 2 | 1 |

B. Sections are subdivided into 16 parts called legal subdivisions, numbered 1 to 16 starting in the southeast corner of the section and ending in the northeast. Each legal subdivision in a regular section is 40 acres.

LEGAL SUBDIVISION NUMBER 7

A SECTION OF LAND: 640 ACRES

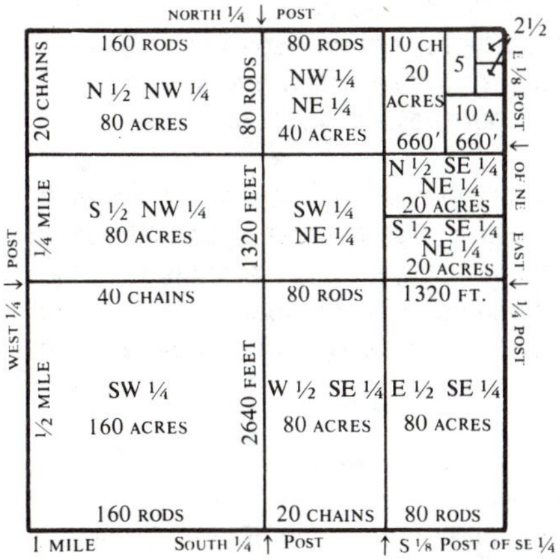

LONG MEASURE

1 mile = 80 chains
     = 320 rods
     = 5280 feet

1 chain = 4 rods
      = 66 feet
      = 100 links

1 rod = 5½ yards
    = 16½ feet
    = 25 links

1 link = ⅔ foot
     = 7⅞ inches

SQUARE MEASURE

1 sq. mile = regular section
         = 640 acres

1 acre = 10 sq. chains
     = 160 sq. rods
     = 43,560 sq.feet

An acre is about 208¾ feet square
An acre is about 8 rods wide, 20 rods long, or any area the product of whose length by its width (in rods) is 160 or in chains is 10.

1 sq. rod = 30¼ sq. yards
        = 272¼ sq. feet

1 sq. foot = 144 sq. inches

C. SUBDIVISIONS OF THE SECTION AND COMMONLY USED UNITS OF MEASUREMENT.

## Figure 2-8.

Divisions and Subdivisions of the Section.

not used as base maps for actual field mapping, topographic maps should be utilized as supplementary reference maps for general orientation purposes. They provide an overview of the mapping area and its relationships to adjacent areas.

For some selected quadrangles in the Topographic Map Series, *orthophotomaps* are produced. Orthophotomaps illustrate natural features primarily by color-enhanced photographic images which have been processed to show detail in true position. They may or may not include contours. Because photo imagery depicts an area in a more true-to-life manner than the conventional line map, orthophotomaps show extensive areas of sand, marsh, or flat agricultural areas and provide additional control points.

*Orthophotoquads* are a basic type of photoimage map prepared in quadrangle map format. They are printed in shades of gray without image enhancement or cartographic symbolization. Orthophotoquads are vertical aerial photos that have been photomechanically corrected for foreshortening due to terrace slopes and yield a planimetrically correct image. Some features are identified and named. The U.S. Geological Survey produces them at the same scale as the standard 7½ minute topographic map. They may serve as excellent resource materials, particularly if used in conjunction with the proper corresponding topographic map (see Figure 2-9). Both the orthophotoquad and the standard 7½ minute topographic map cover the identical area and at the same scale, 1:24000 (about 2⅝ inches = one mile).

The topographic map shows many features that the orthophotoquad does not, such as precise elevations and the areal outline of topographical features utilizing contour lines, bench marks, and spot elevations. The map also indicates the political boundaries of counties, townships, sections, and incorporated areas not present on the orthophotoquad. In addition, the topographic map identifies and names all major natural features, such as rivers and mountains, as well as many cultural features, such as roads, towns, schools, trailer parks, golf courses, and the like. The orthophotoquad, on the other hand, identifies and names only the most prominent features, but it does show many landscape features not on the topographic map, such as field patterns, vegetation types, general soil patterns, farm lanes and minor farm buildings, settlement patterns within urban areas, and minor topographic features with such low local relief that they do not "catch" a contour line.

Together the topographic map and orthophotoquad provide a great deal of information about a local area and also indicate a large number of control points and lines that may be useful for orientation and mapping certain phenomena in the field. Providing the scale of the map and orthophotoquad is compatible for a given field problem, they can serve as excellent base maps as well as supplementary sources of pertinent information.

### Plat (Cadastral) Maps

*Plat (cadastral)* maps illustrate the precise boundaries of a parcel of land. They are generally used to define ownership of property and real estate divisions for assessing and taxing purposes, and for detailed architectural planning. The scales of these maps are large, ranging from 1:600 to 1:10,000. The dimensions of the individual parcels of land or lots normally are given in feet, and the compilation of the square footage of each piece of property is easily calculated (see Figure 2-10). There are no control or reference points other than lot boundaries, which may or may not be visible in the field. For that reason, they do not serve as suitable base maps in rural areas or where extensive units of property exist. However, for certain problems and where lots are small and the land is built up or developed, such as mapping urban land use, they may be the most desirable base map. They are available in each county court house and may be duplicated cheaply. They may be used in conjunction with aerial photos or topographic maps to accurately ascertain farm boundaries and other property lines.

*Sanborn maps** are another type of cadastral map that may be highly suitable as base maps, particularly for urban field research. They, like the conventional plat maps, provide information concerning lot shapes and dimensions, block and lot numbers, street identification, and house addresses. In addition, the buildings and structures are shown as well as their dominant use and a detailed description of the type of construction. These maps are often used to provide information for fire insurance purposes because they contain highly detailed information about the internal structures of buildings and types of fire protection devices.

Most towns and cities have Sanborn map coverage at a very large scale (normally 1 inch = 100 feet or 1 inch = 200 feet), and the maps provide valuable data in urban land-use surveys. The maps are revised periodically, and comparisons of maps at different dates reveal the nature and speed of urban land-use changes.

## Other Base Maps

In specific areas, other types of mapping may serve as suitable base maps. For example, geodetic maps, other than the topographic maps discussed previously, may be useful for certain types of field problems. Planimetric maps produced by many state and federal agencies for areas under their control may also serve as excellent base maps. Agencies such as the U.S. Forest Service, Bureau of Land Management, Bureau of Reclamation, and Soil Conservation Service are but a few examples. Such mapping is valuable in large study areas for exploratory investigations, reconnaissance sampling, and parcelle mapping, to be discussed in Chapter 5. Frequently, these maps are at a scale of approximately one-half inch to the mile and clearly show townships with section dimensions, roads, streams, and built-up areas.

Remote sensing imagery from airborne and other nonsatellite platforms may be used as base maps in the field in certain instances. Aerial photographs were one of the first remote sensing methods to be developed and used widely. Aerial photos translate visible light reflected from the earth's surface into a black-and-white or color image. In recent years sophisticated instruments have been developed to detect electromagnetic energy invisible to the human senses. This information is processed to present a color or black-and-white image. These instruments, usually installed on aircraft or satellites, can record data from considerable distances and provide information about nonvisible objects, features, and phenomena that otherwise would go unnoticed. Data collected by remote sensing techniques are of great value to those scientists concerned with natural and artificial environments.

---

*Produced by the Sanborn Map Company, Inc., Pelham, New York.

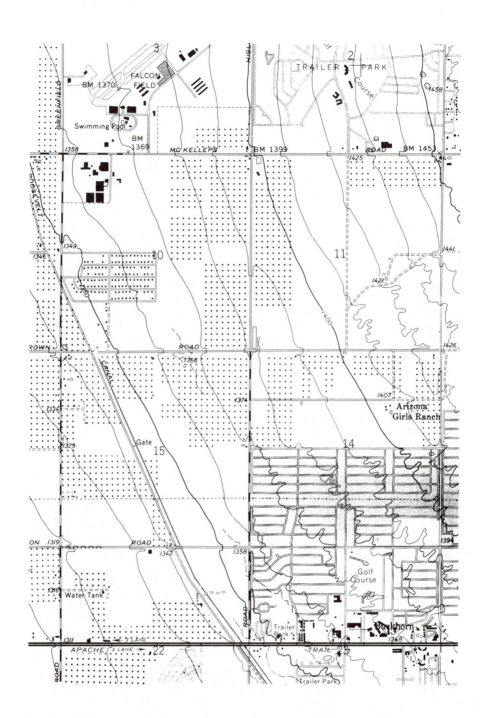

42

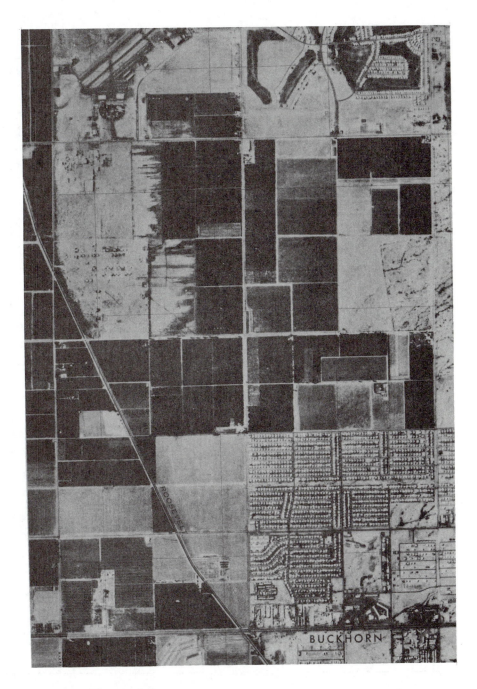

**Figure 2-9.**

A Portion of the Standard 7½ Minute Topographic Map and Corresponding Orthophotoquad. Buckhorn, Arizona.

Both map and orthophotoquad are the same scale. Each illustrates certain features of the natural and cultural landscape that the other does not. Together they provide a great number of control points and lines.

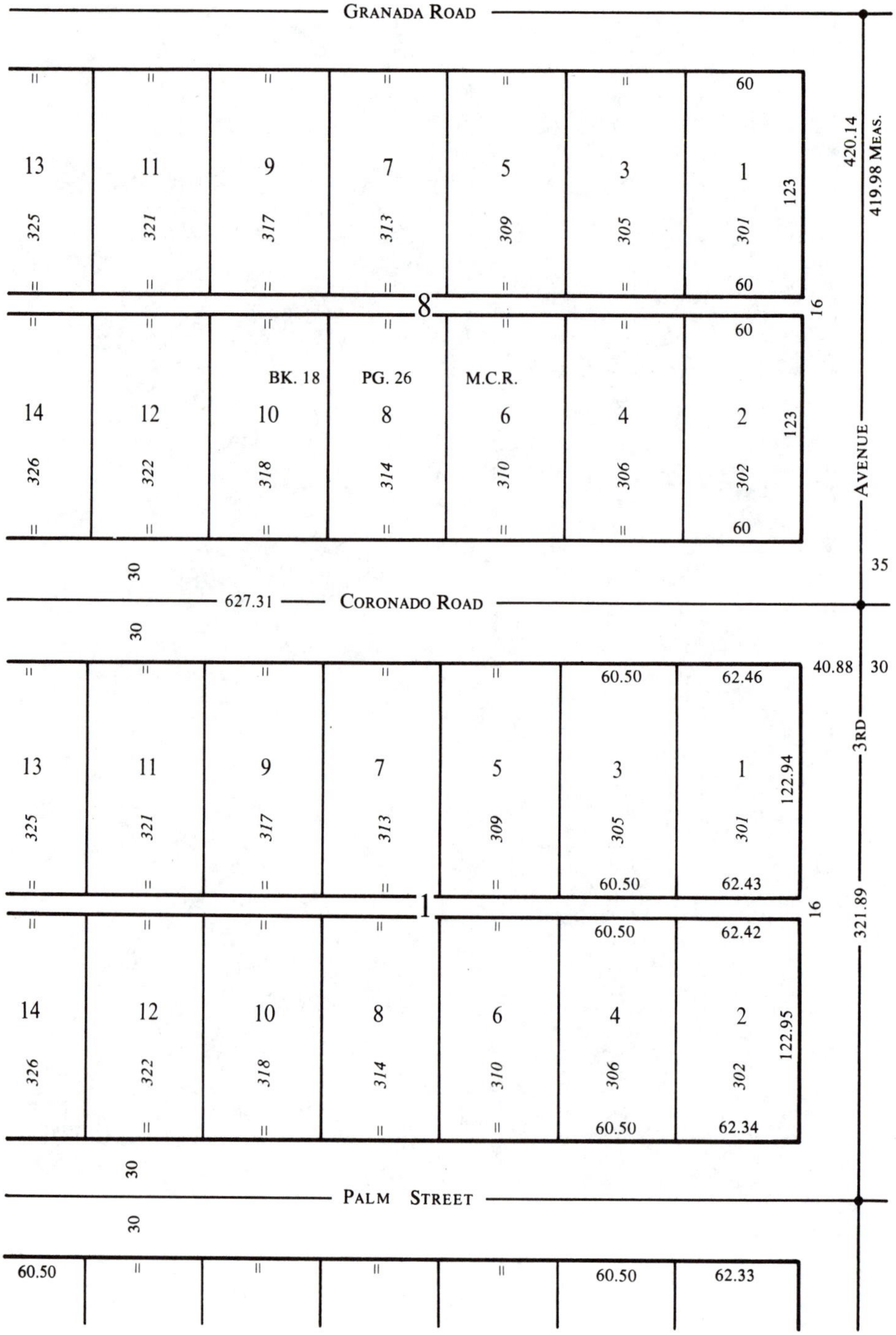

GRANADA ROAD

| 13 | 11 | 9 | 7 | 5 | 3 | 1 |
| 325 | 321 | 317 | 313 | 309 | 305 | 301 |

60
123
420.14
419.98 MEAS.

8

| 14 | 12 | 10 | 8 | 6 | 4 | 2 |
| 326 | 322 | 318 | 314 | 310 | 306 | 302 |

BK. 18   PG. 26   M.C.R.

60
123

16

AVENUE

30

35

627.31 — CORONADO ROAD

30

| 13 | 11 | 9 | 7 | 5 | 3 | 1 |
| 325 | 321 | 317 | 313 | 309 | 305 | 301 |

60.50   62.46
40.88   30
122.94
60.50   62.43

3RD

1

| 14 | 12 | 10 | 8 | 6 | 4 | 2 |
| 326 | 322 | 318 | 314 | 310 | 306 | 302 |

60.50   62.42
122.95
60.50   62.34

16
321.89

30

PALM STREET

30

| 60.50 | | | | | 60.50 | 62.33 |

Figure 2-10.

Sample of a Conventional Plat Map.

44

In general, the nonvisible data acquired is electromagnetic energy radiated (emitted) from objects above absolute zero (O K) or energy reflected from objects or features (sunlight, laser light, or microwave radar). This directly emitted radiation or reflected energy can be translated into images that may be analyzed to distinguish water temperatures, geothermal areas, forest fires, and so on. The reflection of both visible and nonvisible near infrared light may be detected using color infrared film. Infrared film provides identification information (a special signature) to distinguish vegetation types and land use. Reflected laser light and microwave radar may also record relief and topographical features and other surface and subsurface characteristics.

A multispectral survey may be taken using several sensor systems, each designed to measure reflectance in a selected bond of frequencies. These various images of the same area may be combined to produce a composite or a false color single image. Essentially, this is the system utilized by the Landsat (formerly ERTS) satellite multispectral scanner and the newer thermatic mapper that operate from an altitude of over 600 miles. While often invaluable for regional scale research activities, satellite platform imagery is generally unsuited for base maps in the field.

These techniques generate a tremendous variety and amount of data. At this stage of development, most data are for very large areas, and detailed information for small areas is not widely available. The imagery is valuable since it provides a great deal of information suitable for laboratory analysis. At this time, however, the small scale presentation and high cost per frame limits its utility in any environment outside of the office or laboratory.

In all cases, the field researcher should determine what map coverage

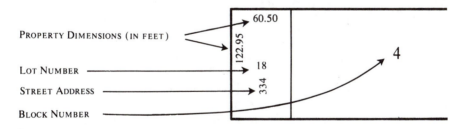

## Figure 2-10 Modification.

An example of a plat map (modified from a portion of the plat) (SW1/4, Sec. 32, T.2N, R.3E., Phoenix, Arizona. *Scale: 1 inch = 100 feet*). Plat maps may serve as excellent base maps for problems dealing with urban land use. Property dimensions are exact and serve as the basis for assessment and taxation. Plat maps of built-up areas are normally available in scales of 1 inch = 50 feet, 1 inch = 100 feet, and 1 inch = 200 feet.

In rural areas where lots or properties are large in size, plat maps are normally not suitable for use as base maps due to the lack of sufficient control or reference points.

exists for the specific research area and carefully evaluate its potential use in terms of scale, date, accuracy, and control points before fully formulating the actual design of the research problem.

## Plane-Table Maps

If no suitable base maps of the field research area exist at a scale desired for field-mapping purposes, accurate base maps may be made by constructing a series of known points and lines, each in its proper direction and distance from the others. These known reference points and lines then may serve as the framework within which specific phenomena may be accurately recorded. This *plane-table mapping* is not normally used for field problems at the intermediate or macroscale since other base maps are usually available. However, in microscale field studies, plane-table mapping may be necessary to obtain the accuracy and precision desired.

In constructing a plane-table map, it is essential that the direction, and the distance from one point to another, be determined accurately. In addition, specific problems may require data concerned with slope or topography. In those cases, differences in elevations from one point to another will also be necessary. (Plane-table mapping is discussed in detail in Chapter 3 and Appendix B.) When a sufficient number of reference points and lines have been established, there exists a controlling framework within which specific phenomena or landscape features may be accurately mapped (see Figure 2-11).

## Measurement of Area

In most field research problems, it becomes necessary to measure or calculate the area (in acres, hectares, square feet, square meters, etc.) of a given phenomenon recorded on the base map. For example, this might be the amount or percentage of land devoted to agriculture or urban uses, or how much of the research area is in a given slope category.

Any phenomenon whose areal extent has been recorded on the base map is subject to areal measurement. There are three standard methods used for this purpose. Which of the three is best depends upon the nature of the specific problem, the scale of the data obtained, the extent or size of the research area, and the intensity or complexity of the information recorded on the base map. The *planimeter* is a mechanical integrator that records linear distance on a graduated drum. The stylus of the planimeter is moved along the boundaries of the area to be measured, and the linear distance recorded can then be calibrated to a square measure.

The second method is to cut the areas to be measured from the base maps with a knife, scissors, or other instrument and weigh the accumulated pieces on a balance scale. Prior calculations will determine the ratio between the weight of the paper of the base map and the square or area measure. This procedure, called the *cut and weigh* method, usually is employed when the research area is large and the phenomena to be measured are highly detailed and irregular in areal extent. This method obviously destroys the base map, so what other uses that were planned for the original field base map should be completed prior to the area measurements.

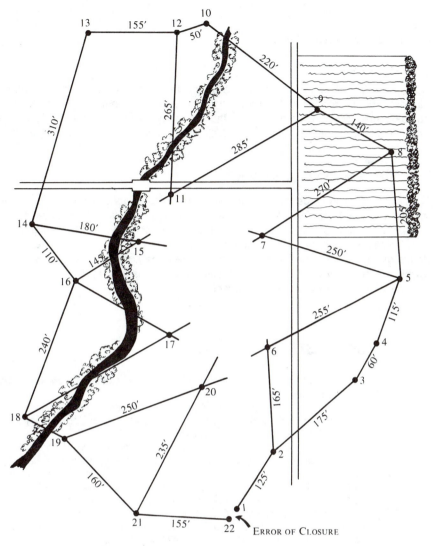

Figure 2-11.

Sample Plane—Table Map (*Scale: 1 inch = 160 feet*).

A series of control points has been established as
well as a line of traverse (closed traverse). From
this framework, other control or reference points
may be located accurately, and the areal extent of landscape features such as land use, vegetation, slope, soils, settlement, etc., mapped.
From each station of observation, the exact direction and distance to other stations must be
determined if the base map is to be accurate. If
the last sighting does not coincide precisely with
the first station, the gap is known as the error of
closure. An error of closure may be due to errors in sighting or in measuring distance (particularly if paced or tape measured over irregular
topography), or to miscalculations in arithmetic. A
gap of 3 percent or less of the total length of
the traverse is acceptable (see Appendix B).

The third method employs a transparent *template* with a grid or random dot network for calculating the areal unit of measure. The template is placed over the base map, and the number of units that cover the area of the phenomenon to be measured are counted. With some modifications, this method can be adapted to computer use.

## Base Map Selection

The selection of the most appropriate base map(s) for a specific problem must be made with great care. The advantages, limitations, and special features of each type of map considered must be weighed and evaluated critically. The type(s) of base map(s) selected will, to a large degree, set the pattern of subsequent research in terms of the accuracy of the data to be collected, the speed and efficiency of the mapping, the analysis of the data collected, and the manner in which the conclusions and results of the study are presented.

# SUGGESTED REFERENCES

BIRCH, T. W., *Maps: Topographical and Statistical.* London: Oxford University Press, 1964.

DICKINSON, G. C., *Maps and Air Photographs.* London: Edward Arnold, 1969.

DOWNS, R. M. and D. STEA, *Maps in Mind: Reflections on Cognitive Mapping.* New York: Harper & Row, Publishers, 1977.

ESTES, J. E. and L. W. SENGER, eds., *Remote Sensing: Techniques for Environmental Analysis.* Santa Barbara, California: Hamilton Publishing Co., 1974.

GREENHOOD, D., *Mapping.* Chicago: University of Chicago Press, 1964.

HOLZ, R. K., ed., *The Surveillant Science: Remote Sensing of the Environment.* Boston: Houghton Mifflin Co., 1973.

LOW, J. W., *Plane Table Mapping.* New York: Harper & Row, Publishers, 1952.

*MANUAL OF PHOTOGRAPHIC INTERPRETATION.* Washington, D.C.: American Society of Photogrammetry, 1960.

*MAP READING,* United States Department of the Army, Technical Manual FM 21-26, 1969.

MILLER, V. C. and C. F. MILLER, *Photogeology.* New York: McGraw-Hill Book Co., 1961.

MONKHOUSE, F. J. and H. R. WILKINSON, *Maps and Diagrams.* New York: E. P. Dutton & Co., 1963.

PUGH, J. C., *Surveying for Field Scientists.* Pittsburgh, Pennsylvania: University of Pittsburgh Press, 1975.

SPEAK, P. and A. H. C. CARTER, *Map Reading and Interpretation.* London: Longmans Group Ltd., 1977.

# Instruments, Measurements, and Mapping Techniques Used to Acquire Information From Observable Features and Objects

## Chapter

# 3

**Chapter**

**3**

Many types of measurements, mapping techniques. and instruments are commonly employed to gather data about visible landscapes. These include a wide range of instrumentation, from highly complicated and specialized instruments such as the theodolite (employed for surveying) and radiometers (used for measuring energy balance) to such simple devices as the pocket stereoscope, pocket magnifiers, shovels, thermometers, and the common pencil. In fact, simple observational techniques of data gathering by estimation and mapping without use of instruments are by themselves effective in certain field studies. Some of them are quite accurate and, within the scope of their intended utility, rival or exceed the ease with which accurate data may be obtained using field instruments.

## MEASUREMENT SCALES

As mentioned in Chapter 2, there is a world-wide trend to employ metric measurements whenever possible. To be effective, however, the field researcher must have a working knowledge of both the metric scales and their English equivalents and be prepared to use whichever is compatible with the area and time of research.

There are many measurements, particularly those associated with cadastral surveys, that are not readily changed into understandable metric equivalents. For example, townships and sections are more conveniently discussed in the English units from which they were derived. It should also be noted that many base maps, particularly those in the topographic quadrangle map series, which are available for most areas of the U.S., presently use English units. The U.S. Geological Survey is planning a new series of topographic maps at a scale of 1:100,000. Even the contour intervals in this new series will be metric. However, until this series becomes available, the existing English-based topographic maps of the United States must be used.

Often, a field researcher must use base maps that employ units other than English or metric. For example, Spanish land grants used the vara unit discussed in Chapter 2. Old survey documents employed the chain-and-link measurements. In addition, all deed records are expressed in English units, a format not likely to be altered in the near future. It is therefore necessary either to adapt to the units being used on the base map and cadastral records, or to convert all measurement data to the metric scale. The latter is not an easy task, and the degree to which it is feasible depends on cost and time constraints.

Most data other than cadastral information may be converted to metric units without undue difficulty. In fact, much of the instrumentation used today is calibrated in metric units. Temperature data, for example, are gathered in Celsius degrees as well as in Fahrenheit. Energy, soil, vegetation, and wildlife measurements have long used the metric scale as a standard.

Some common units of measurement stand by themselves. For example, the units used for horizontal and vertical angles are based on the division of the circle into 360 degrees, with each degree subdivided into 60 minutes, and each minute subdivided into 60 seconds. This universally employed system of measure dates back to Babylonian times.

## MEASUREMENT ACCURACY

A discourse on instrumentation of measurements would not be complete without considering accuracy. Accuracy is a function of several variables:

1. Instrument error,
2. Operator error,
3. Existing error, such as may be present in cadastral documents dating back to ground control errors of previous surveys,
4. The degree to which the field researcher is precise in acquiring any quantitative or qualitative data.

This last point is extremely important since time spent in the field may be translated into dollar costs, and there are definite constraints with regard to the level of precision that is cost-effective. Some methodologies and analytical techniques do not require high levels of accuracy for their application, while quantitative and statistical analysis generally require the highest orders of accuracy. In some field studies, particularly at small scales and in exploratory and reconnaissance work, lower levels of precision are acceptable.

51

The field researcher must also guard against collecting falsely precise data. For example, if stream channel parameters such as width, depth, and bed angle are being collected, the method employed to determine the limits of the channel for measurement must be carefully considered. If the existing water level is used as a reference, measurements to the nearest centimeter might be in order; however, if the field researcher is confronted with a dry wash or low water flow, some other criterion would have to be used. Perhaps only an "educated guess" could fix the limits of the channel, and measurements to the nearest decimeter or even nearest meter might be the closest that realistically could be employed.

The correct use of *estimation* can be a very valuable technique in field data gathering. In measuring plant cover or amount of land devoted to commercial land use, for example, a percentage estimate is frequently used, recognizing that the percent estimated by a particular individual may differ from that of another. As long as one person makes all of the measurements, the fact that they are all high or all low is not of vital concern. Relative to one another, they are correct. Some field mapping methodologies are based on estimation where the relative values are the important measures. If several individuals are involved in a field study, they all must employ similar percentage value estimates. Again, it is not a serious matter whether the entire group's estimates are high or low, just as long as they are consistent. Estimation, together with many simple qualitative measurement techniques, is applicable to and complements the more rigorous instrumented measuring procedures.

## RESOURCE MEASUREMENT AND INSTRUMENTATION

Resource measurements and instruments, useful to geographic field work, may be classified into five broad and general groups. These are:

1. Cadastral and topographic measurements concerned with direction, distance, and differences in elevations. These measurements are most important in that they accurately locate points and lines, which individually and collectively serve as the framework within which a great variety of phenomena can be observed and recorded.
2. Measurements concerned with surface or near-surface materials such as surface textures, soil types, and rock outcrops. These measurements are normally associated with pedological and related studies and are of concern only to specialists in these fields.
3. Measures concerned with surface cover such as natural vegetation, land use, and other types of cover. These measurements include a wide variety of phenomena and are of concern to most geographers, both physical and cultural.
4. Measurements dealing with hydrological or meteorological phenomena concerned with specialized studies such as microclimatology, and surface and subterranean drainage.
5. Space adjustment measurements concerned with phenomena in motion such as traffic, wildlife migrations, and movement of commodities. These measurements are pertinent to a wide range of geographical studies and are of interest to most geographers.

## Cadastral and Topographic
## Measurements and Instruments

Location of phenomena is basic to geographic field work. The dimensions of direction and distance from one place to others are essential to the description and spatial interpretation of all objects, features, activities, and events. They are common to all geographic problems, although the degree of precision necessary to determine them will vary from one study to another, depending upon the scale of the research. For this reason, the bulk of this discussion concerning measurements and instrumentation focuses on cadastral and topographic measurement. Cadastral or planimetric mapping includes techniques and instrumentation used to ascertain physical and cultural shapes and patterns and their relationships to the surface configuration of the landscape. Topographic measurement is concerned with the determination of the surface configuration itself. Similar instruments and techniques are used for both types of measurement. A representative list of instruments used for such measurements include the plane table, open and telescopic alidades, levels, transits, theodolites, various compasses, altimeters, tapes, chains, level rods, and range rods.

**The Plane Table and Open-Sight Alidades.** In most field work today, adequate base map coverage exists in the form of aerial photography, planimetric coverage, or topographic coverage. In microlevel studies, however, existing photography or regional mapping sets frequently do not cover the study area in sufficient detail to be used as base maps. In these instances, researchers must create a base map in the field prior to other data gathering. This can be done easily, and in some instances cheaply, by the use of the *plane table* and an *open-sight* or a *telescopic alidade* (see Figure 3-1).

For most field mapping, a simple plane table and alidade are preferable to the more sophisticated transit or theodolite since the mapping is done on-site using the table itself. With theodolites and transits, written field notes are made, but the actual mapping is done back in the office. Most field researchers without surveying experience are not prepared to make proper field notes. Furthermore, most field studies or projects do not have the cartographic support necessary to translate these notes into a finished map. Therefore, plane tabling is the less difficult since the map can be checked in the field while it is being drawn.

The plane table consists of a drawing board mounted on a tripod. In its simplest form, the drawing board is permanently fixed to the tripod and no angular tilting adjustments are allowed. Such a table, called a *traverse table*, is built for speed of operation rather than for high accuracy (see Figure 3-1). Normally, the traverse table is used in conjunction with open-sight alidades rather than with more complex instruments. The open-sight alidade is a simple sighting device that basically consists of a ruler with peep-sight vanes attached (see Figure 3-2). A common triangular ruler or scale may be used when no alidade is available.

The more sophisticated drawing board and tripod combinations contain a leveling head and clamp, which allow for both rotation and tilting (see Figure 3-1). Using a table that can be leveled precisely is necessary for accurate mapping. The most elaborate table mounts have three leveling screws in addition to the rotational adjustments. The *Johnson table*, as an example, has two setscrews fixed to a central, dual-threaded shaft (see Figure 3-1). The lower setscrew controls the

Instruments, Measurements, and Mapping Techniques

TRAVERSE TABLE

TILT     AZIMUTH

JOHNSON TABLE

**Figure 3-1.**

Plane Tables and Instruments.

The traverse table is used with open-sight alidades
for planimetric maps of low accuracy. The Johnson table with telescopic alidades is used for planimetric and topographic mapping of high accuracy.

rotation of the table; the upper setscrew controls the tilt of the table. The Johnson
table is normally used with telescopic alidades (see Figure 3-2). Frequently, the
edge of the drawing board is inletted and equipped with a small compass needle
and housing to assist in orienting the table. The drawing surfaces of the plane tables
have four to six flat-head setscrews placed in shallow depressions along the periph-
ery of the table so that the drawing paper can be drawn tightly against the surface
for accurate work.

The equipment necessary for plane-table work, in addition to the plane
table and tripod, includes an open-sight or a telescopic alidade, a scale (preferably
an engineers' scale for measuring distances), a sharp, hard pencil (harder than
4-H), a long straight pin, drawing paper, and a sighting rod. The paper used
depends on weather conditions and length of time anticipated for the plane-table
survey. It may be conventional drafting material or frosted acetate. Frosted acetate
or mylar may be used in wet weather since the markings will not rub off as they
would on conventional paper.

**The Telescopic Alidade.** If vertical as well as horizontal control
is required, the telescopic alidade must be used. This instrument consists of a
telescope mounted on a steel straightedge (see Figure 3-2). The telescope is
mounted on a horizontal axis which allows it to be pointed up or down through an
arc of about 50 to 60 degrees. The eyepiece of the telescope is at the extreme rear

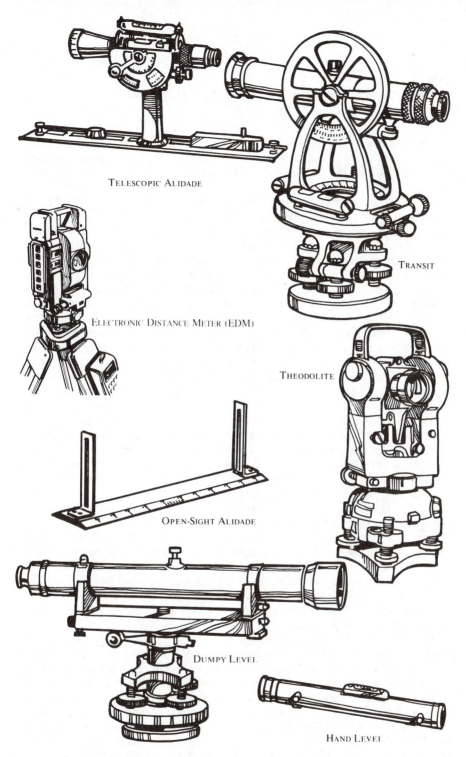

TELESCOPIC ALIDADE

TRANSIT

ELECTRONIC DISTANCE METER (EDM)

THEODOLITE

OPEN-SIGHT ALIDADE

DUMPY LEVEL

HAND LEVEL

## Figure 3-2.

Mapping Instruments.

Of the variety of mapping and surveying instruments shown, the hand level and the open-sight and telescopic alidades are those most frequently used by field geographers.

and may be one of two types: (1) a conventional direct-viewing eyepiece, in which case the user looks through the back of the telescope; or (2) a prism eyepiece, which enables the user to look through the telescope from the top or from either side.

The focusing knob is on the right side, midway down the barrel of the telescope ("right" and "left," for purpose of discussion, are determined by viewing the instrument from the rear or eyepiece position). An additional focusing adjustment, consisting of a knurled ring located at the rear of the barrel, is used to focus the cross hairs. The correct procedure for cross-hair focusing is to point the telescope away from the sun, look at the open sky, and, by rotating the ring, bring the cross hairs into sharp focus. If the user attempts to focus the cross hairs while viewing ground objects, the hairs will not be in focus for all distance settings of the main objective lens.

The telescopic alidade also is equipped with a circular bubble along the baseplate which is used to level the table. On top of the instrument is the primary leveling bubble, called the *striding level*, which is loosely attached to the top of the telescope. It is not clamped to avoid distortion of its mount. The *Beaman arc level* is located on the left side of the telescope. One of its uses is to ascertain the alignment of the instrument. Associated with this level, located beneath the main axis of support, is the Beaman stadia arc and vernier scale, calibrated in degrees. This is used to measure vertical angles of objects as well as to calculate stadia corrections for large angles of view upwards or downwards. In addition, when both the Beaman arc bubble and the striding bubble are exactly level, readings of 30° and 50° are opposite each other on the vernier scale. This alignment indicates that the instrument is in adjustment and has not been damaged.

On the right side of the telescope, and slightly to the rear of the vertical mount, is the *Stebinger screw*, which is used to slowly elevate or depress the telescope. A circular ring mounted on the screw is calibrated from 1 to 10 and is used to determine elevations when a sighting is either above or below the level rod position. In addition, a compass is located to the rear and left side of the instrument.

Some telescopic alidades are set on a short pedestal directly on the top of the base. Others, like the one shown in Figure 3-2, are much taller and have 12–15 centimeter pedestals. The smaller version is more useful for reconnaissance and movement in rough terrain.

A view through the telescope reveals several cross hairs. The vertical cross hair is used for sighting on the range pole so that mapping rays may be drawn in the correct direction (see Figure 3-3). The longest horizontal cross hair marks the level position of the instrument and is used in reading elevations from the rod. A pair of markings called stadia marks are positioned above and below the horizontal cross hair. The footage difference in the rod readings between these smaller cross hairs is called the *stadia interval*. This interval multiplied by 100 is equal to the distance from the instrument to the rod. For example, the stadia interval shown in Figure 3-3 is 1.3 feet, so the distance between the instrument and the rod is 130 feet ($1.3 \times 100 = 130$).

Several types of level rods are in general use. One of these is the *Philadelphia rod*, a wooden pole with a metal or plastic face calibrated in feet, tenths, and hundredths of a foot, with zero at the base of the rod (see Figure 3-4).

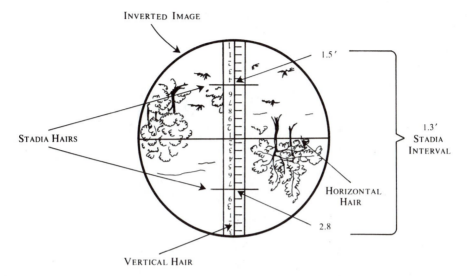

**Figure 3-3.**

A View through the Telescopic Alidade.

Depending on the instrument used, either an
erect or inverted image is seen. Note the four cross
hairs and their relationship to the level rod. Measurements are in feet.

The rod may be extended to a height of 14 feet, and when fully extended, it is called
*high rod.* Another form of level rod is called the *Linker* or *self-reading rod.* The
calibrations are quite different, and numbers increase going down the rod. A movable metal tape is fixed to the face of this rod and allows an actual reference
elevation to be set on the telescope cross hair. When the rod is viewed in another
location, the actual elevation of that new point is read from the rod. The Philadelphia
rod does not read elevations directly. The Linker rod is of greatest utility in low-relief
terrain where elevation differences of about six to eight feet exist. Setting the Linker
rod is time consuming, especially in rough terrain; therefore, it is often easier to use
the Philadelphia rod, which does not require setting and extends to a higher position
at high rod than the Linker.

        **Other Leveling and Related Survey Instruments.** The tele-
scopic alidade is the surveying instrument most commonly used in geographic field
work. There are, however, several others which find lesser use in field geography.
These include the dumpy level, the hand level, the transit, and the theodolite (see
Figure 3-2).

        *The Dumpy Level:* The *dumpy level* consists of a surveying telescope and
support assembly mounted on a tripod. There is no plane table or drafting surface.
Four leveling screws and a bubble level assembly are attached to the telescope. The
same leveling procedures and elevation determination as employed for the telescopic alidade are used. However, field notes instead of a map are generated. Such

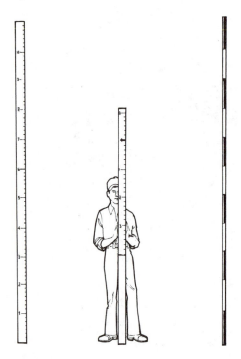

Figure 3-4.

Types of Level and Sight Rods.

These principal types of rods are used in plane-table mapping. The Philadelphia rod has graduations increasing upwards from zero at ground level. The Linker or self-reading rod has a movable face with numbers that increase toward the ground. The line or sighting rod is a small-diameter pole that is painted, not numbered, to indicate even foot markings.

levels find many applications outside of map making, such as grade height determinations on construction projects.

*The Hand Level:* For many purposes, a *hand level* is sufficiently accurate and convenient in the field. There are two forms of hand levels. One consists of a bubble tube mounted on top of the telescope; the other has a split mirror device that allows the user to see the position of the bubble in relationship to a horizontal reference mark while looking through the telescope. When using the hand level, the observer must stand erect and firmly hold the instrument to maintain a constant height. Proper use of a hand level will enable the user to produce fairly accurate contour maps of small areas and to carry rough elevation differences from one point to another. For example, along a stream bed, the effective sight distance is up to about 50 feet, and with care, reasonably accurate elevations may be carried for 400 to 500 feet, within the accuracy range of approximately one foot.

*The Transit and the Theodolite:* The *transit* is a precision instrument that is used principally to measure horizontal angles and to extend straight lines over long distances. It may be used for stadia measurement as well as leveling and is frequently termed the universal survey instrument. The *theodolite* is similar to the transit in many respects. It is used for still-higher levels of precision surveying and astronomical observation. The principal differences between the two instruments are the length of the telescope and the fact that the telescope of the transit can be rotated about its horizontal axis and pointed forward or backward (see Figure 3-2). The level can be placed on top of the transit telescope for level work. Back-sights are easily made by flipping the telescope over. The theodolite telescope is much shorter and is not mounted high enough to allow rotation about its horizontal axis, thus allowing for greater accuracy. Four leveling screws are located on a leveling plate on both the transit and the theodolite.

In addition to the optical instrument an electronic theodolite or Electronic Distance Meter (EDM) has a built in microcomputer and measures distances using a modulated infrared or visible light laser beam. Instruments like these can calculate distances, heights of objects, and elevations, and perform triangulations and other calculations. The digital output of these instruments may be recorded and subsequently plotted in map or profile form on microcomputers. They represent the highest order accuracy surveying instruments. All types of theodolites and transits are used in geographic field work only when highly precise measurements are necessary. They are the tools of the surveyor rather than of the geographer.

*The Altimeter:* Elevations also may be determined with the surveying *altimeter* (see Figure 3-5). Although less accurate than the instruments described previously, the altimeter is frequently used and will yield acceptable results as long as proper monitoring procedures are employed. The surveying altimeter is a precision aneroid barometer that translates barometric (air) pressure into altitude. The American Paulin System surveying altimeters are accurate within five feet, and in some cases, within two feet. Temperature compensation calculations, as well as calculations to account for diurnal barometric pressure change, should be made. The surveying altimeter first must be placed at a known elevation point, and the instrument set to this reading. The temperature must be recorded, and a barograph should be placed at the initial bench-mark station to record diurnal pressure changes (see Figure 3-9). The altimeter may then be carried from place to place and read as required. The typical survey altimeters are manually set to a zero reading for each observation. Every few hours, the altimeter should be returned to the original bench mark or known elevation. It is likely that temperature and pressure changes will have occurred, and correction calculations need to be made for all measurements previously taken. The user must return frequently to the original known elevation to assure that observations are being made only during times of constant pressure change.

Less sophisticated altimeters do not require temperature or pressure corrections and are merely set to the proper elevation and carried about during the field work. These instruments are not highly accurate and may be several hundred feet in error depending on the quality of the instrument and the existing conditions.

*The Compass:* The *compass* is used widely for cadastral measurement. There are several types of compasses: the *Brunton pocket transit*, used for surveying; the *Lensatic compass*; and the *Silva* or *ranger compass* (see Figure 3-5). The

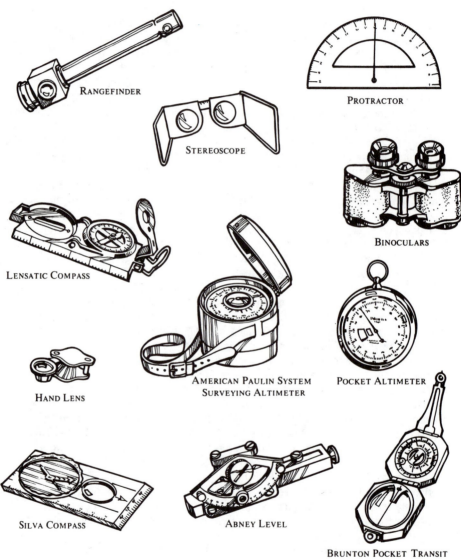

RANGEFINDER

PROTRACTOR

STEREOSCOPE

BINOCULARS

LENSATIC COMPASS

HAND LENS

AMERICAN PAULIN SYSTEM
SURVEYING ALTIMETER

POCKET ALTIMETER

SILVA COMPASS

ABNEY LEVEL

BRUNTON POCKET TRANSIT

## Figure 3-5.

Small Pocket or Portable Instruments.

These instruments are in widespread, common use
by field researchers.

60

Brunton compass may be used for direction calculations or as a level, clinometer, plumb, or alidade. Like the transit, it may be used for taking vertical or horizontal angles. The graduated compass scale is divided into either four 90° quadrants for surveying, or into a 360° counterclockwise azimuth circle for use primarily in land navigation. For field geographers, the 360° azimuth circle is the more useful since there is less possibility of error. All types of compasses have a magnetized needle suspended on a pivot (or needle with magnets attached) that orients to the earth's magnetic field; a pivot protector, which lifts the needle off of the pivot pin when the case is closed; and some form of sighting device, both front and rear (except for the Silva compass). Lensatic and similar types of compasses can be adjusted to compensate for magnetic deviation. The Lensatic compass has a small lens set into the rear sight. This lens allows the user to lower his or her eye while taking a sight and see the magnified azimuth scale beneath the *lubber* or zero line on the compass. By shifting the gaze downward from the horizontal, a compass reading may be made. When using any type of compass, great care must be taken to avoid metal objects, including metal on the person of the user to offset magnetic attraction. Very simple compasses, when properly used, can record accuracies within one degree over distances of more than a mile. The principal use of the compass in geographic field work is in locating sample points and moving accurately and efficiently from one known point to another.

The Silva or ranger compass is a fluid-dampened device in which the liquid reduces needle oscillations. The azimuth scale and fluid container are mounted on the top of a plastic plate with tenths of an inch and millimeter scales calibrated on the side. The container and its azimuth scale may be rotated to set a particular compass heading into the instrument. A parallel series of lines, with arrows showing the direction of north, rotates with the container and is visible by looking down through the top of the compass. The user merely turns the entire compass and plate assembly until the needle lines up with these parallel lines. Sighting may then be accomplished by looking over the top of the compass. The Silva compass is widely used because of its simplicity and low cost.

Slope orientation is determined by taking a compass bearing along the line of greatest slope of the surface. When measuring slope, the problem of scale or level of resolution must be considered. Microsite variations in slope exist over distances of a few meters, while in macrosites (areas of a hundred meters or more), more generalized trends in slope may be observed. Beyond this macroscale, there exist regional trends into which an entire study area may fall. The choice of scale is, of course, determined by the size of the research area and scope of a given problem. Questions of scale must also be resolved when considering relief classifications and other forms of surficial configuration documentation. Frequently the pocket or full-size stereoscope and appropriate aerial photography are useful in making such decisions (see Figure 3-5).

*Abney Level:* The *Abney level* is an instrument used to determine vertical angles in the field (see Figure 3-5). It commonly is graduated from 0° to 90° with a 60° vernier but may also be calibrated in percent slope or in topographic units. Similar to the hand level, the Abney level is used by marking a point on the rod that is level with the user's eye height and held by a second individual 50 feet or less from the instrument position. The user turns a knob on the left side of the instrument to center the bubble, and the slope angle may be read directly from the vernier.

An even simpler device, consisting of nothing more than a protractor and a paperclip, is also commonly used to determine vertical angles. The paperclip is straightened, except for a small hook at one end which fits through the center hole in the flat side of the protractor. The straight wire is allowed to hang free against the graduations along the curved edge, and the user sights along the flat side, much in the same manner as when using the Abney level. The degree of slope may be read from the side by another person, or the protractor tilted carefully to maintain the paperclip at the proper degree marking. This technique should be repeated several times to insure accurate readings. Over short distances, it is accurate to about a degree. One advantage of using a protractor instead of the Abney level is that the slope of surfaces may be measured from a distance. The flat side of the protractor is held parallel to the slope, and the angle read directly from the instrument.

Often, the least complex instruments are the most useful in the field. Therefore, instruments should not be disregarded out of hand because of their lack of sophistication. For most field work, measurements from simple instruments will produce acceptable results unless, of course, high orders of accuracy are required. In many geographic field studies today, time-conserving and cost-effective methods which yield reasonable data are carried on and prove to be valuable for the purposes intended. Similarly, simple sketch mapping, field sketching, and estimation of distances using pacing and other methods often yield useful information, particularly in exploratory and reconnaissance situations.

*Distance Measures:* A variety of methods are available for determining horizontal distance. The stadia rod and alidade methods have already been discussed. Other methods include the use of chains, cloth and metal tapes, special measuring wheels, automobile odometers, pedometers, and pacing (see Figure 3-6).

*Short Cloth and Steel Tapes:* Short cloth and steel tapes of 50 feet or less are portable and useful for certain purposes. In making measurments, the zero end of the tape is carried away from the reference point or station. On most tapes, the zero point is the center of the wire forming a loop at the end of the tape. This is placed over the point that is to be measured and the distance is read and recorded at the reference point by the individual holding the tape case. For best results, the tape must be held horizontally and stretched slightly; with cloth tapes a minimum amount of pressure should be applied, but for steel tapes a firm pull provides the most accurate measurement.

*The Chain:* Distances greater than the length of small tapes are best measured with a chain. The chain is not actually made up of links (although early forms of surveying chains were formed by links, with each link a definite measure). The surveying chain is a spring steel strip with lead patches on which foot markings are inscribed. Chains are available in 100-and 200-foot lengths. The 100-foot length is the most convenient to count points or stations over long distances. Steel chains have a leader on the head end, or zero part, of the chain. This consists of an additional one-foot distance subdivided into tenths of a foot. Therefore, the full

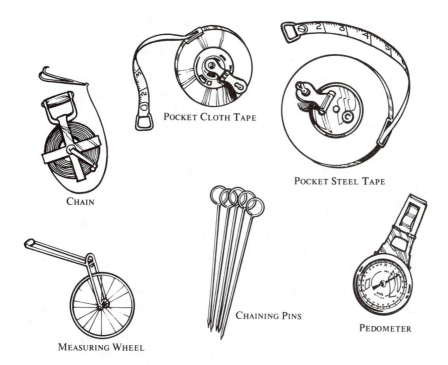

**Figure 3-6.**

Distance-Measuring Devices.

length of the chain from leader to tail is 101 feet. Again, two individuals are necessary for proper use of the chain.*

Over terrain that is undulating, level chaining is necessary if accuracy requirements are high. Bubble levels and plumb bobs are used together with scales and levers to pull the chain to specified tensions. Depending on the accuracy required, someone may have to look up the coefficient of expansion of the chain and apply temperature corrections. Distances may be in error in excess of 6 to 8 feet in

---

* It is common practice to use *chaining pins* or arrow points. There are eleven pins to a set, and the individual heading up the chain takes the zero end with ten pins. The eleventh is left behind with the tail of the chain, either marking the station or held by the tail chainperson. The head chainperson places one pin every 100 feet, and as the tail chainperson walks forward, each pin is picked up, and a tally of the exact number of 100-foot stations equals the number of pins at the tail. Every 1000 feet, the tail and head trade pins and the tail chainperson starts with one pin again. For long distances, this is the most efficient way to keep track of the 100-foot stations (the number of 100-foot segments of the total distance measured).

When using the chain the person at the head always places the zero mark on the point or station to which the measurement is going. The person tailing the chain pulls it taut and reads to the nearest foot; if tenths of foot measurements are required, then the closest foot is held by the tail chainperson and the excess, tenths of feet, are read directly from the leader portion of the chain. Therefore, if the closest foot marker is 47 feet and the distance is somewhere over 47 feet, then the 47-foot mark is held by the person tailing the chain and the tenths are read by the person heading the chain. Employing standardized procedures is the safest method to avoid error. The most common mistakes are a one-foot error, a 100-foot error, or a tenth of a foot error, any of which could be serious.

10-mile traverses when extreme temperatures exist between times of two comparative surveys.

*Optical Distance Measuring:* Optical distance-measuring devices are frequently useful in the field (see Figure 3-5). These, often called range finders, are particularly advantageous during reconnaissance and exploration work. Used in conjunction with a good set of binoculars, they can reduce the amount of time spent in walking from one part of the field area to another by allowing enlarged inspection of selected areas to determine which field subsites need personal visitation.

## Surface and Near-Surface
## Measurements and Instruments

Pedological (soil) measurements are usually taken in highly specialized field studies. However, even in general field studies, soil types are frequently checked against existing Soil Conservation Service mapping to establish a data base for a variety of geographic field problems.

It rarely is necessary to identify and completely analyze the soil types in an unknown environment. Usually some soil mapping exists, which, at a minimum, documents the range of types that might be encountered.

Since it is necessary to dig into the ground to view horizons, soil structure, color, and other visible features, the basic soil instrument is the shovel (see Figure 3-7). The shovel should be of the "sharp-shooter" type, with a straight, narrow, and long blade, for smooth-sided, deep holes. A surplus military entrenching tool is useful since it functions as both a hoe and a shovel; however, the short handle makes it a poor choice for deep holes. A metal hand trowel is also helpful for collecting samples for laboratory analysis. In soil that is not stony, augers are frequently used; some are equipped with a collection tube for taking core samples.

Soil samples are traditionally gathered at the surface, in representative zones of the various soil horizons, and close to the bedrock level. It is necessary to have soil sampling bags if material is to be removed for later laboratory analysis. Plastic "zip-lock" bags are the most convenient. However, soil stored in such bags must be processed quickly since bacterial action in the closed environment may change the pH and other chemical conditions of the soil. If soil must be stored for a long time prior to analysis, paperbags are best because the soil will dry out rapidly. Sample volumes should be around 240 milliliters (about a cup) to be adequate for laboratory analysis.

Frequently, on-site field determination is the most convenient method of soil analysis. Field pH meters and analysis kits to test for calcium, phosphorus, and other chemicals are available. Soil moisture content may be measured with small inexpensive meters pushed into the ground. These instruments are marginally accurate and should be used only in soils that have a moderate moisture content.

Field estimation of soil texture is frequently employed. Researchers must educate their fingertips to the feel of various soil textures, for example, silty clays, clay loams, sandy loams, sandy clays, clays, sands, and so on. This can be done by preparing jars of mixtures of sand, silt, and clay that conform to the conventional standard texture categories. The accepted field method is to put a small amount of soil in the palm of the hand and then spit into it. Roll the wet soil

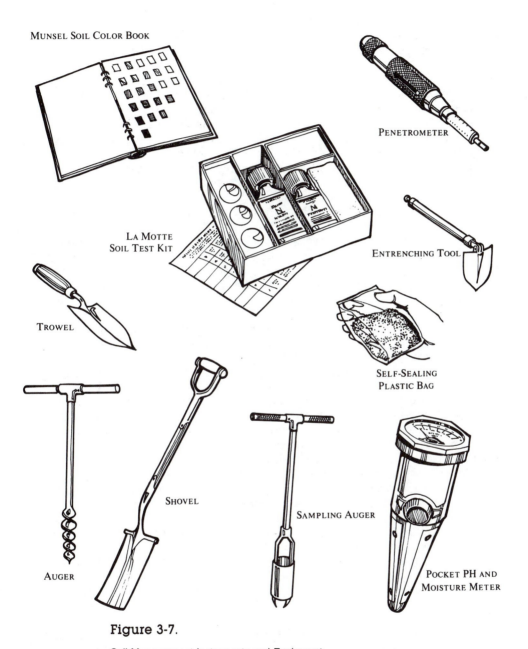

**Figure 3-7.**

Soil Measurement Instruments and Equipment.

into a ball with the fingers. If the soil contains high quantities of clay, it can be rolled into a stringy shape; clay soil will also tend to feel slick or sticky between the fingers. Lack of clay or high silt content prevents the soil from staying together. Quantities of sand may be felt easily since the grains will tend to "cut" the fingertips. A loam soil may be squeezed and pressed easily. It will not stay together and will act much like a sponge.

Another technique that is used to estimate stoniness of a soil horizon uses the common pocketknife. A knife is stuck, at random, deeply into the sides of a soil pit. This procedure is repeated ten times. The number of "hits" (stones) that can be felt in the handle of the knife represents the percentage of gravel. For example, if three stones are encountered in ten tries, the stoniness is 30 percent. Although crude, this technique does provide a relative measure of the percentage of stones or gravel in a soil. Another crude, but useful, measurement of soil moisture content is made by placing a soil sample in a small plastic bag and sealing the top. If the bag is laid in the sun, any moisture within the soil will vaporize from the heat and condense along the inside of the bag. The amount of moisture that condenses in a certain length of time is a realtive measure of the moisture content. More sophisticated soil moisture measurements employ conductivity blocks placed at various levels in the soil. Soil material may also be weighed in both wet and dry conditions.

Field measurements of color are made easily using the Munsell soil color books. Because high soil moisture content affects the color of a soil, a dry soil color measure is the best in many instances. Color measurement, of course, can be done in the laboratory.

In addition, density measures for soil surfaces are often necessary. An instrument called a *penetrometer* can be used to determine, for example, whether or not trampling by animals has occurred in certain vegetation areas.

In some instances, soil covering is lacking and bedrock is exposed at the surface. The type of rock, dip, and strike of inclined strata are of interest to the physical geographer as well as the geologist. The critical measurements of *dip*, or angle of inclination, and direction of inclination may be determined by using the Brunton pocket transit previously discussed. The dip is measured by a clinometer, which consists of a pointer and level bubble. The pointer, attached to the inside of the compass, can be moved against a scale graduated in degrees. The compass is held in a position paralleling the dip slope, or tilt, and when the bubble is leveled, the angle of dip is read directly from the graduated scale. The direction of dip is expressed in terms of compass direction. *Strike* is the compass direction of the line of intersection of the tilting strata with a horizontal plane. It is always perpendicular to the direction of the dip and can be determined by using a Brunton or several other types of compasses.

The percentage of rock outcrop can be determined by using the ccver estimation techniques discussed in the following section. Rock or mineral identification in the field implies some knowledge of mineralogy and petrology. Identification keys are available and sample specimens can be brought back to the laboratory for detailed analysis.

Several remote sensing techniques may be employed at ground level to gather microsite data from below the ground surface. Ground penetration radar, proton magnatometers, and soils resistance measurement devices may be used to detect areas of soil compaction, near surface moisture relationships, subsurface cavities, archaeological sites, and urban data such as location of buried pipes and utilities. The sensed depth varies from one to several meters depending on site conditions.

## Vegetative Cover and Land-Use Measurements and Instruments

The way that the surface of the land is used is of major concern to geographers. Settlement patterns, rural and urban land use, and natural covers are common subjects studied in the field. Frequently, the land use reflects the topographic and pedologic conditions discussed previously.

**Natural Vegetation.** In most instances, the field geographer is involved only in limited vegetation sampling. Primarily, he or she makes estimates of various vegetation types and plant parameters such as cover, frequency, density, vigor, yield, height, and sociability. These estimates can be complemented by the appropriate soil measurements to determine how a particular vegetation stand may be classified. Similarly, data concerned with agricultural crop-type classification and estimates of vigor, yield, and stage of maturity may be collected.

The techniques for measuring and estimating any cover type or land-use category, both rural and urban, are similar in many ways. Rarely will the geographer be involved in determining the types of vegetation within an area that has not been previously studied. Most frequently, the task in the field is to match published information against what is actually found in the field. Therefore, estimations of all sorts are quite useful.

The most effective estimations are made by using categories rather than continuous scales. For example, when estimating percentage covers, the ranges 1–10 percent, 10–25 percent, 50–75 percent, and 75–100 percent should be used. Estimating a specific percentage, say 46 percent, is probably not possible anyway.

Aerial photography of a suitable scale could be employed to determine percentage of ground cover or land-use type. This measurement may be done in the laboratory as well as in the field. To estimate vegetative cover, the researcher might observe the area in question and through a combined visual and mental process, place all plants of the same species together, and ascertain what portion of the plot they cover by mentally dividing the area in half, and halves into quarters, and asking the question, "Is more than a quarter or more than half of the plot covered by a particular species?" Through this technique, the researcher can get repetitive, comparable measures of plant cover.

Another estimation method is to count the phenomena being mapped to determine a particular item relative to an area or a series of plots. For instance, if an item occurs in 10 out of 40 plots, it has a "frequency" of 25 percent. If there are 50 items total in the 10 plots, then the "density" is 5 per unit (plot).

Other measurements depend on the subject matter and are easily made after the researcher becomes familiar with the appropriate literature and gains field experience. For vegetation, examples would include vigor, yield, and sociability.

Height is frequently measured initially with a meter stick or metric tape to mentally fix the general height ranges of objects and is estimated thereafter.

A camera is a valuable vegetation field-sampling instrument. It has the capability of recording all that it sees. Comparing photographs taken in the field with photographs in published materials is an excellent way of studying phenomena.

In determining species, the researcher should take samples of the plants back into the laboratory and consult a specialist unless he or she is skilled in taxonomy. Certain species are quite difficult to identify and, in various parts of the United States, published field keys are not adequate for proper identification, unless the researcher has a great deal of experience.

Other instruments and equipment are also useful during field vegetation analysis. The most commonly used instrument is the stereo-dissecting microscope and associated dissecting kit. This instrument is used for field determination of grass species. A variety of pocket magnifiers are also valuable for identifying vegetation (see Figure 3-5). For those specimens that cannot be identified in the field, or for herbarium collections and documentation of species, a *vasculum* is useful. The vasculum is a canister carried on the belt to hold whole specimens or parts of specimens in a humid environment. A plant press and associated pads are also required if the distance between the field site and office or living quarters is great (see Figure 3-8). Even in a vasculum, plants will eventually dry out to a condition where they are not suitable for pressing.

Munsell color books may be used in the field to determine flower and leaf color. In forested environments, the *increment borer* is used to remove a core from the center of a tree for counting the annual growth rings and determining the growth pattern over the tree's lifetime. Similarly, *calipers* and other devices for measuring trunk diameter and tree height may be employed (see Figure 3-8).

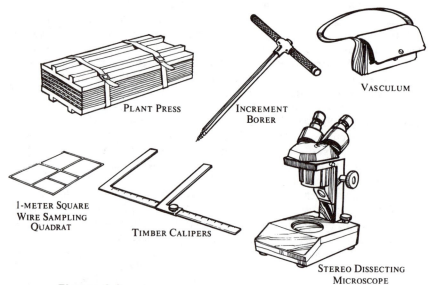

PLANT PRESS

INCREMENT BORER

VASCULUM

1-METER SQUARE WIRE SAMPLING QUADRAT

TIMBER CALIPERS

STEREO DISSECTING MICROSCOPE

**Figure 3-8.**

Equipment for Vegetation Sampling.

These are the most common types of equipment for this special form of field work.

**Land Use.** There are no special instruments to measure land use that have not been discussed previously. Cadastral measurements are employed to determine land-use boundaries and locations of specific features. Aerial photographs are frequently employed to record rural and urban land uses, and in high-density urban areas, plat maps are used. The major problems in measuring land use focus on identification and classification. The great variety of possible land uses necessitates devising a system of grouping in order that measurements may be taken. The most frequently used systems of land-use classification subdivide rural land uses into broad categories such as cultivated land, pasture, forest, grassland shrub, and idle land. Cultivated land may be further subdivided into the type of crop produced; pastureland into permanent, rotation, woodland, or brushland pasture; forest land and grasslands or shrubs into categories reflecting the dominant species; and idle land into fallow, abandoned, or wasteland categories. Similarly, urban land uses can be grouped into major general categories such as residential, commercial, industrial, institutional, governmental, and vacant. Residential land use may be subdivided into single-family, multiple-family, or apartment; commercial land use may be divided into specific types of uses such as furniture stores, supermarkets, hardware stores, etc., or kinds of commercial districts such as shopping centers, isolated store clusters, etc.; and industrial land uses may be subdivided into categories such as textile plants, food processing, printing and publishing, etc. Governmental land uses can be subdivided into waterworks, fire stations, court houses, parks and recreation, etc.; and institutional land use may be further categorized into schools, churches, hospitals, etc.

Presently there is no universal or standardized land-use classification system. Although efforts have been made toward standardization in the United States for over a decade, each locality develops the system that best suits its needs. Relevant information concerning land use includes the actual activity on or use of the land, the physical characteristics and potentialities/capabilities of the land, economic potentialities/capabilities of the land, value, ownership, and environmental quality. In developing a workable land-use classification system, the following problems must be resolved:

1. The selection of land-use phenomena to be recorded;
2. The scale or minimum area to be mapped or recorded;
3. The development of a system of categorization;
4. Inclusion of nonvisible characteristics such as value and age of structures, socioeconomic characteristics of the population, etc.;
5. Rotational land uses—short-term and temporary changes that fit into a consistent pattern; and
6. Mixed land use revolving around the determination of the dominant use or combinations of repeatedly occurring land uses classified as a specific category.

The development of a land-use classification system is most critical since the system used determines how the data may be compiled, analyzed, and presented.

## Meteorological and Hydrological Measurements and Instruments

Most field studies concerned with meteorological and hydrological phenomena are highly specialized. For this reason, only a brief description of the more commonly used instruments will be given. Most of the instruments mentioned are used for general monitoring purposes rather than for sophisticated research studies.

**Meterological Measurements.** A great variety of field instrumentation exists for meteorological determinations (see Figure 3-9). Measurements include precipitation, temperature, pressure, dew point, visibility, radiation, wind speed, humidity, and sky dome. Instrumentation that has a recording capability of one week or even longer can be used for most of these values, allowing one individual to monitor several sites simultaneously. Note that we are talking about meteorological measurement rather than climatological measurement here. Climatological data gathering takes place over long periods of time, and the type of field work that is being discussed here is of short duration, involving one or several visits rather than twenty years or more of record taking.

Precipitation instruments include conventional rain gages, recording or tipping-bucket rain gages, and weighing gages for rain or snow. The last two record intensity of precipitation. Air temperature, surface soil temperature, and subsurface temperature, as well as temperature conditions aloft, may be measured with a variety of thermometers, thermocouples, and telemetry equipment. Aneroid barometers find wide field application in barographs and precision altimeters. Sometimes, even the highly accurate and calibrated mercury barometer is used in the field.

Dew point may be measured with dew-point hydrometers, whirled psychrometers, sling psychrometers, or aspiration psychrometers. Hair hygrometers are often incorporated with barographs or with thermographs. Such an instrument is known as a hygrobarograph, or hygrothermograph. Hand-held anemometers, of either the Robinson cup type or the arrow vane type, are employed to measure wind velocity. Energy is monitored through the use of radiometers. Radiometers are incorporated into elaborate radiation instrumentation sets to measure heat flux and energy budgets at particular field sites (see Figure 3-10). The sky dome, the 360° horizon levels for a particular site, is also measured. This is accomplished through the use of a Brunton compass or similar instrument, and azimuth and elevations of the horizon are taken and plotted for a 360° arc.

**Hydrologic Measurements.** A second major category of physical resource measurement devices is concerned with hydrologic data gathering. A variety of measurements are commonly made by geographers using hand-held portable equipment. Stream velocity is measured with small flow meters which look like miniature anemometers and may be placed at chosen depths by positioning the instrument on a support rod. Readings are possible down to 0.1 foot per second (see Figure 3-10).

Water surface-level measurements are made using the Gurley hook gage, a device which permits readings to 0.1 millimeter in still water over a depth of about 6 decimeters. In addition to water flow, water quality is frequently of interest and includes observations of suspended sediment load, channel bed material, chemical composition, dissolved oxygen content, and concentrations of pollutants. Suspended sediment is measured with rod-mounted samplers. These, as well as the flow instruments, are commonly used while wading, from low bridges, or from small boats. Each sampler contains a collection bottle which fills with water and sediment as it is lowered and raised at a uniform rate from the water surface to the stream bottom. Chemical and pollutant concentrations are measured with field analysis kits specifically designed for a range of chemical tests.

## Space Adjustment Measurements (phenomena in motion)

A wide variety of phenomena that might be measured in the field are in motion, constantly adjusting the spatial relationships of the landscape upon which they are operating. These include the following:

1. Movements of wildlife;
2. Air currents;
3. Stream flows,
4. Vehicular traffic on all types of roads and highways;
5. Movements of goods by ships, aircraft, rail, and common carrier; and
6. Human movements without mechanized assistance.

Measurement of any of these phenomena presents the field researcher with a large and common problem: An individual can only be in one place at any given time. For meaningful measurement, simultaneous observations at multiple locations are required. One solution is the use of automatic mechanical counting devices that may be left in several locations and collected after the measurement period. Examples include traffic counters, recording stream-flow gages, recording anemometers, etc. Radio telemetry equipment is used extensively for certain wildlife migration monitoring. Selected animals are tagged with radio transmitters, and their movements, as well as those of the herd, are followed with direction-finding equipment. Information about biological functions are also telemetered; thus, feeding habits and sleeping locations may be recorded. The principal drawback to the use of telemetry is its high cost.

Alternatively, a large number of human observers, equipped with measuring instruments or hand-operated counters and stationed at strategic locations, may be used. This is an inexpensive way to gather data on traffic flow, for example. If a map of morning rush-hour traffic flow within a central business district is required, the personnel and equipment might be as follows:

1. Four observers on each major thoroughfare intersection (one on each corner to cover all possible traffic-flow directions), and two on minor intersections. If the central business district (CBD) had 5 major and 25 minor intersections, 70 persons would be required.
2. A number of mechanical counters to tabulate adequately the flow routes of cars. Since each traffic stream could either go straight, turn left, or turn right, three counters per corner would be required, or a total 360 for the entire CDB.
3. At least one timepiece per intersection to determine start and stop times accurately. Thus, 30 watches would be required.
4. Finally, since the recording task might be hectic at rush hour, some spare individuals for relief work.

Over longer periods, such a data-gathering procedure quickly becomes impractical.

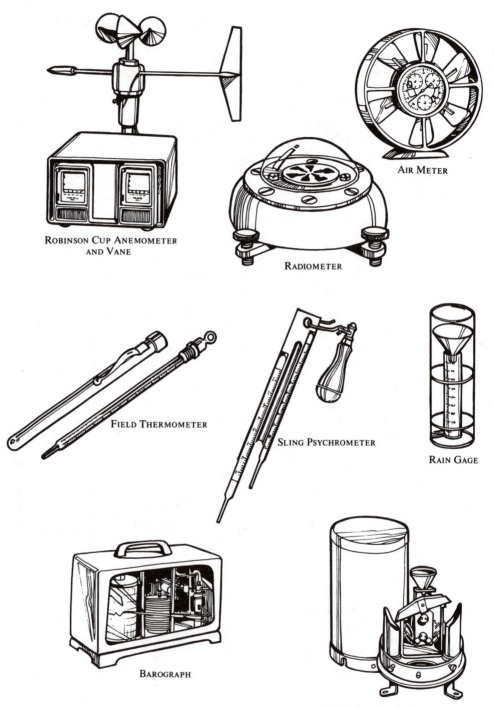

ROBINSON CUP ANEMOMETER
AND VANE

AIR METER

RADIOMETER

FIELD THERMOMETER

SLING PSYCHROMETER

RAIN GAGE

BAROGRAPH

TIPPING BUCKET RAIN GAGE

72

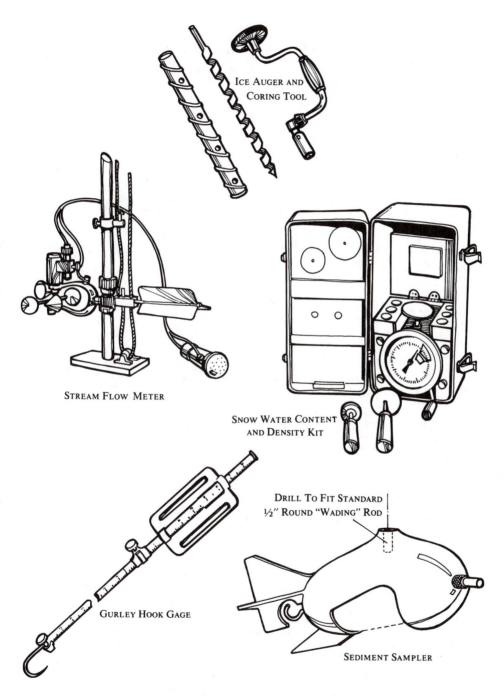

ICE AUGER AND
CORING TOOL

STREAM FLOW METER

SNOW WATER CONTENT
AND DENSITY KIT

GURLEY HOOK GAGE

DRILL TO FIT STANDARD
½″ ROUND "WADING" ROD

SEDIMENT SAMPLER

◄Figure 3-9.▲

Meteorological and Hydrological Instruments.

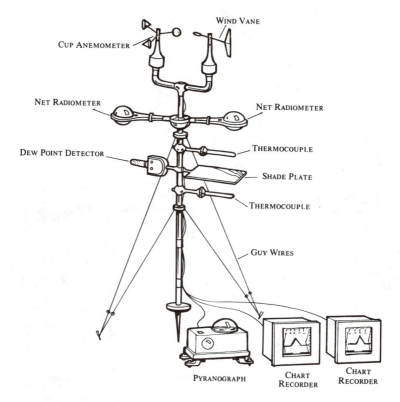

CUP ANEMOMETER

WIND VANE

NET RADIOMETER

NET RADIOMETER

DEW POINT DETECTOR

THERMOCOUPLE

SHADE PLATE

THERMOCOUPLE

GUY WIRES

PYRANOGRAPH

CHART RECORDER

CHART RECORDER

**Figure 3-10.**

Energy Budget Field Station.

This setup includes devices for monitoring radiation, temperature, humidity, and wind motion.

## COMPUTERIZED DATA ENTRY AND TELEMETRY

The compilation of all forms of field generated data is an important activity. Traditionally data are gathered, compiled, and used in nondigital formats. Then at some later time, depending upon project requirements, the researcher's philosophical approach, and other constraints, some or all may be digitally encoded and processed for the project.

For extensive field surveys portable microcomputers are becoming an invaluable tool. Data may be entered immediately on site and edited on-line. Maps and tables become instantly available for proofing and managing data gathering. Some disciplines already make extensive use of microcomputers. Archaeological surveys are a notable example; the coordinates of objects uncovered within a dig, the artifact description, and other data are entered as the articles are excavated. The data bank is being built continually and used for map and daily work activity summaries. This way of handling data is just beginning to be employed in geography and the biological sciences.

One pervasive problem with the use of microcomputers in the field is the environment of the field itself. Several serious problems related to climate must be addressed: excessive heat or cold, very high or low humidity, dust, physical shock, and insects can all disable a computer system. Electrical power is also an important consideration: a clean, stable, uniform voltage, and reliable source are absolute necessities. All these obstacles can be overcome by physically sealing the microcomputers, constructing modifications for cooling, and using sealed hard disk drives and audio tape storage systems instead of floppy disk drives that are easily dust damaged and cannot be effectively sealed. Systems are also available that deliver the necessary clean electrical power. Protecting against physical shock and rough handling is largely a matter of careful moving and selecting suitably rugged hardware.

Some limitations can be overcome by using small hand-held units (Hewlett Packard and others) with internal random access memory (RAM) of sufficient size to permit a usable data bank to be entered; or by using radio-telemetry to a central site (van or camp) from the field mapping location. This central site could be better protected from the environment and standard microcomputers could be used without extensive modifications. Off-the-shelf hand held transceivers with built in keyboards (Motorola 12DX 1000) are rugged and immune to most environmental problems. They have the added advantage of adding quality communications to a field endeavor for efficiency and safety.

An activity closely related to recording primary field data is environmental monitoring over long time periods. Presently various recording devices (both analog and digital) are employed for these activities. Syncronization and event timing are not easily coordinated between measuring sites when using discrete devices, but this problem can be solved if all data are telemetered either over phone lines or by radio into a central data gathering site. This practice is standard in industry.

Survey instruments and questionnaires generated by computer allow interviews to be efficient and almost error free. Most large marketing research firms have employed this technology for the past decade. Recent advances in microcomputers and portable battery operated equipment will make similar procedures usable even in field situations.

## MAPPING TECHNIQUES

To insure that the field data collected are meaningful and adequate for further analysis, the specific mapping techniques employed must be appropriate to the given field problem. There are conventional and widely employed techniques that suffice in most situations, but modifications to and innovations on these should be considered and, if workable, the amended techniques should be utilized. The overriding objective of field mapping is to acquire the essential information in the most accurate and efficient manner possible. The specific overall field problem, including its components of scale and classification systems employed, limitations of the base maps used, and whether or not the total research area is to be mapped or only selected sample areas, determines what mapping techniques should be devised and used.

Several mechanisms of measurement discussed previously, such as estimation, also may be considered to be mapping techniques. It is not always possible, nor is it a matter of vital concern, to make a distinction. Measurements and mapping techniques are closely interrelated. The mapping techniques described in the following discussion are representative, not exhaustive, and are selected to demonstrate a wide range of ways that observations may be mapped in the field.

### Point Locations and Isolines

It is not uncommon in many mapping problems to find that the base map does not have an adequate number of reference points. In this event, additional *point locations* can be determined by triangulation, intersection, and resection (see Figure 3-11). If two points are known, a third can be located by determining its direction from each of the two known points and plotting these bearings on the map. Additional point locations and associated base lines can be added to the base map to construct a controlling framework sufficient for a given problem. Either a good compass and topographic map or a plane table may be used for this.

A procedure known as resection is used to locate phenomena of interest while in the field, and also to ascertain the exact location of the researcher within the spatial matrix. Compass bearings are taken *to* known objects that are visible both on the ground and on the map. Next, a bearing *from* the known objects is obtained to plot the observer's position. To do this, 180 degrees are either added or subtracted from the *to* bearing to obtain a reciprocal, or *back azimuth*. The intersection of the back azimuth rays establishes the observer's position.

If values are assigned to point locations, it is possible to construct an isogramic map. An *isogram* or *isoline* is a line connecting points of equal value. The values may be elevation, resulting in a contour map, or other physical, economic, or cultural phenomena. Isogramic maps may show the spatial distributions of attitudes or perceptions, densities of population or traffic flow, time-distance ratios, or any phenomenon subject to measurement and value determination. These maps can be made in the field at the time of mapping or in the laboratory at a later date. The accuracy of such maps depends largely on the number or density of point locations that are mapped and assigned values.

### Single-Feature and Multiple-Feature Mapping

*Single-feature* mapping is concerned with determining the areal extent of one aspect of the environment, such as slope, land use, or house types, etc. Single-feature mapping often will suffice for a given problem if other required data are available from existing sources. Multiple-feature mapping implies recording information about two or more aspects of the area. The number of phenomena that can be mapped in this manner is huge, but the complexity of the field mapping and subsequent analysis increases with the number of landscape features mapped. If careful preparation and planning are exercised, several features can be mapped in one run through the research area. A simple hypothetical example is illustrated in

Figure 3-12. Our problem might be to determine the land use of a rural area and to see if areal associations exist with selected physical characteristics of the land such as slope and the stoniness of the surface. We further assume that if areal associations exist with selected physical characteristics of the land such as slope and the

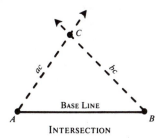

INTERSECTION

A. Points *A* and *B* are known points. After orienting the plane table at point *A*, a line of sight of indefinite length is drawn from point *A* to *C* (line *ac*). A similar sighting is taken from point *B* to *C* (line *bc*). The location of *C* is determined where lines *bc* and *ac* intersect. For the greatest accuracy, the two known points should be located so that the lines of sight intersect at an angle of 60° to 120°.

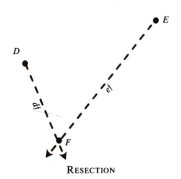

RESECTION

B. The mapper's location is unknown, but the locations of points *D* and *E* are known. By drawing lines of sight back to the mapper's position, the location of point *F* can be determined.

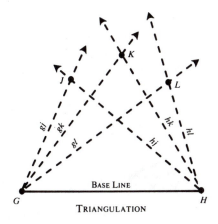

TRIANGULATION

C. Triangulation is several intersections taken from the same base line. Intersection is taken on several points. Points *G* and *H* are known, and by several intersections, the locations of *J*, *K*, and *L* can be determined.

### Figure 3-11.

Intersection, Resection, and Triangulation.

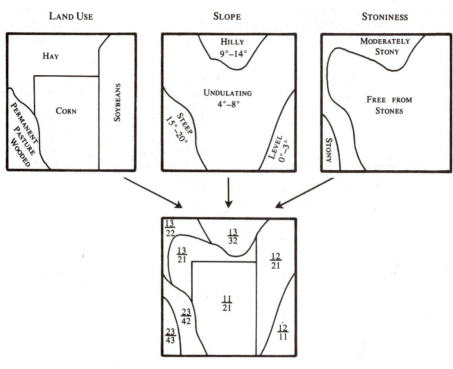

SINGLE-FEATURE

LAND USE            SLOPE            STONINESS

MULTIPLE-FEATURE

## Figure 3-12.

Hypothetical Examples of Single-Feature and Multiple-Feature Mapping.

### GENERALIZED CLASSIFICATION SYSTEMS

RURAL LAND USE: 1st symbol, numerator, general land use; 2nd symbol, numerator, specific land use.

1. *Cropped Land*
   1. corn
   2. soybeans
   3. hay etc.

2. *Pasture*
   1. rotation
   2. permanent, nonwooded
   3. woodland pasture etc.

3. *Forest, Grassland, Shrub* etc.
4. *Idle Land* etc.
5. *Miscellaneous* etc.

SLOPE: 1st symbol, denominator.
1. *0°–3° (level to gently sloping).*
2. *4°–8° (undulating to rolling).*
3. *9°–14° (rolling to hilly).*
4. *15°–20° (hilly to steep slopes).*
5. *21°–30° (steep slopes).*
6. *31° and more (very steep slopes).*

STONINESS: 2nd symbol, denominator.
1. *Free from stones.* Surface sufficiently free from stones so that there are no difficulties in cultivating the land.
2. *Moderately stony.* Small- or large-sized stones in sufficient quantity to impede cultivation or render it difficult.
3. *Stony.* Sufficient quantity of small- or large-sized stones to virtually preclude cultivation.
4. *Bedrock exposure.* Bedrock exposed, or is within plow depth of the surface and covers sufficient space to preclude cultivation.

stoniness of the surface. We further assume that classification systems have been devised previously for each of the three phenomena mapped. A shorthand method of notation may be employed to record our observations on the base map. A widely used system of notation is the fractional code system of mapping. In mapping rural areas, the numerator of the fractional notation refers to the land use, and the denominator represents the physical characteristics of the land (see Figure 3-13). A fractional code system is simply a group of symbols that facilitate mapping in the field. There is no standardized system. A code or system of symbolization should be devised to be compatible with the field problem.

## Sketches and Photographs

In the past, the ability to capture a scene in a field sketch and sort out the relevant details was an important one. Today, however, field sketching is rarely done. The skill has been replaced by more time-efficient methods. The camera has taken the place of the pencil. In one way this is unfortunate, because a camera records everything, while the field sketch is selective and can emphasize the relevant features of the terrain and omit those that are not important.

Initially, landscape sketching was employed to show views from artillery positions within range to describe important targets. Later it was adapted into the geographer's realm of data recording. It is important to think of field sketches and photographs in the present context not as artistic efforts, but as maps in a vertical plane. Their principal purpose is to record information quickly. The present-day use of field photography ranges from pictures made with precision horizontal viewed stereo mapping cameras to simple field snapshots, made to gain an overall geographic perspective or to elaborate on the cultural and/or physical details of a portion of a region. In the modern context, hand-held oblique aerial photography with 35–mm cameras constitutes a valuable supplement to vertical and more conventional aerial photography and to planimetric mapping. It allows areas of interest to be recorded for later study in the laboratory.

Black and white photographs are cheaper and, for most purposes, better than color shots. However, color, color infrared, and black and white infrared film also provide useful information to the field researcher.

The scheduling of field photography coincident with vegetative events such as budding and leaf set and changing of leaf color with the season may aid in plant identification, and even assist in the determination of soil type and some moisture conditions. Polaroid cameras are useful because photographs are instantly available, allowing a check on photograph quality and for note taking directly on the image. A 4 × 5 Speed Graphic camera with a Polaroid back is recommended. In addition, we should not overlook air photography as a data-gathering tool. Different film and filter combinations and stereo pairs are quite useful in certain field endeavors. Also, ground truthing of air photo image components is an important part of geographic field analysis. Here, such image characteristics as tone, texture, pattern, shadow pattern, grain, color, and stereo form need to be ground checked to determine whether or not they result from objects on the surface such as vegetation or cultural items, or from other differences such as soil or rock color. When used in this manner, aerial photographs are particularly cost effective.

A. RURAL.

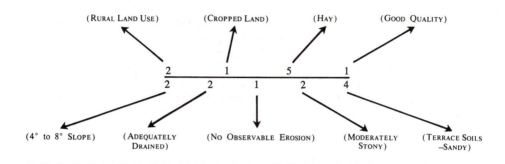

(RURAL LAND USE)　　(CROPPED LAND)　　(HAY)　　(GOOD QUALITY)

$$\frac{2}{2} \quad \frac{1}{2} \quad \frac{5}{1} \quad \frac{1}{2} \quad 4$$

(4° to 8° SLOPE)　(ADEQUATELY DRAINED)　(NO OBSERVABLE EROSION)　(MODERATELY STONY)　(TERRACE SOILS —SANDY)

B. URBAN.

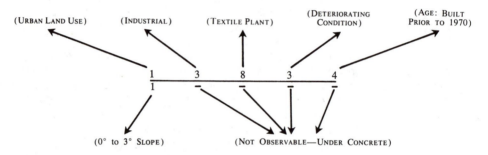

(URBAN LAND USE)　(INDUSTRIAL)　(TEXTILE PLANT)　(DETERIORATING CONDITION)　(AGE: BUILT PRIOR TO 1970)

$$\frac{1}{1} \quad \frac{3}{-} \quad \frac{8}{|} \quad \frac{3}{-} \quad \frac{4}{-}$$

(0° to 3° SLOPE)　(NOT OBSERVABLE—UNDER CONCRETE)

The hypothetical examples above illustrate the recording of land use and quality, and selected physical characteristics of the land (slope, drainage, erosion, stoniness, and soil). Each landscape feature has its own classification system.

## Figure 3-13.

Examples of Multiple-Featured Mapping Using the Fractional Code System.

The use of ground-based telephoto lenses or normal photographs made with fine-grained black and white film allow for detailed inspection of parts of a study area not easily accessible. The film image can be enlarged many times to act as a telephoto view.

In some field situations, it is desirable to collect a series of uncontrolled photographs for landscape mosaics. To do this, the camera is placed on a tripod, and slightly overlapping pictures are made in the horizontal plane. If one horizontal pass does not capture the entire vertical development of the site, the camera angle may have to be elevated and the series repeated. Moderate vertical as well as horizontal overlap should be employed. Later these photographs may be mosaicked, producing an uncontrolled view of the terrain. A mosaic may even be made for a full 360 degrees. This technique is particularly spectacular when employed in mountainous terrain.

With all field photography, systematic documentation of the camera and image variables is useful. These variables include camera location, azimuth of view, type of film, size of lens, filter (if used), "f" stop, and shutter speed. Frequently a scale reference in the form of a meter stick, rule, coin, vehicle, some object of known dimensions, or a person is included in the photograph. Such information will facilitate the use and interpretation of information from the images during the analysis phase of the field project. Such information will also prove valuable to future researchers who wish to rephotograph a certain scene to document landscape changes through time.

## Dimensional Mapping

In certain urban field studies, it may be necessary to obtain land-use or related data for the upper floors of high-rise buildings as well as for the ground-floor levels. Such studies usually are concerned with square-foot analysis of commercial and other urban uses. Delimitation of the central business district, comparisons of different types of commercial districts, and land-use intensity studies are but a few examples. The mapping techniques are not complex and consist essentially of mapping or collecting data from the upper floors of a building at the same scale as from the ground floor.

Most urban land-use mapping is done on plat maps which provide the dimensions (length and width in feet) of the lot or the ground level of a building. The information on the plat map serves as the basis for constructing base maps for each floor to be mapped. The dimensions are transferred to graph paper, usually in block or half-block units, at the scale of the plat map or magnified to whatever scale is desired. Normally the front square footage of the first, and perhaps the second, level is mapped on the ground by circling the entire block. Data for the upper levels must be obtained by direct observations and may be supplemented by information from the building directory. The individual squares on the graph-paper base maps allow for accurate collection and calculation of area data. The result is a series of maps of the entire block, each map representing one floor level. As an alternative to a series of maps, the base map of the various floors may be shown on one piece of graph paper. The profile method consists of constructing a base map for each of the upper floors at the same scale as for the ground floor (see Figure 3-14A). Land-use data are recorded for each floor. Each floor is considered as one story in vertical dimension.

An alternative to the profile method or a series of maps for each floor is to divide the building area in half. Normally a diagonal line is drawn to depict the ground floor half fronting the street, and the remaining half represents second-floor use. If third-floor use is to be recorded, the second-floor area is divided and the third-floor use recorded in the triangle (see Figure 3-14B). Because the uses of floors beyond the second floor frequently are uniform in type, one map may be employed to represent multiple floors.

The technique of using graph paper as a mapping base is also useful in other forms of urban field-data acquisition. Studies involving existing commercial-

A. DIMENSIONAL OR PROFILE MAPPING.

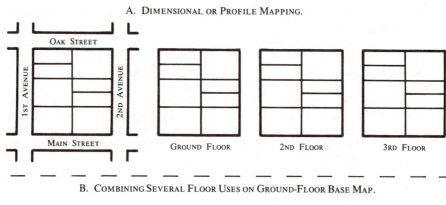

B. COMBINING SEVERAL FLOOR USES ON GROUND-FLOOR BASE MAP.

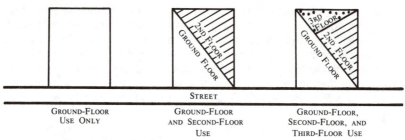

GROUND-FLOOR
USE ONLY

GROUND-FLOOR
AND SECOND-FLOOR
USE

GROUND-FLOOR,
SECOND-FLOOR, AND
THIRD-FLOOR USE

## Figure 3-14.

Mapping Floors of Different Uses.

strip developments, park areas and facilities, industrial parks, and also hotels and small towns may be effectively carried out based on graph-paper maps. Frequently, when plat maps are either unavailable or highly inaccurate, a simple graph-paper map made by pacing and estimation of locations and limits of activities is sufficiently accurate to give the necessary spatial organization and perspective to successfully complete the study. As with all mapping procedures, the physical process involved in locating each store, building, or activity not only makes the researcher intuitively aware of the site of an activity but also tends to give him or her a more complete functional and holistic view of the area, which can be valuable in the analysis phase of the study.

## Sequential Comparisons

The physical and cultural components of the landscape are continually changing. Traffic and pedestrian flows change rapidly and, as we have seen, often present the researcher with measurement problems. Other changes, such as the evolution of landforms, take place over very long periods. In some areas significant alterations in the landscape occur as the result of urban expansion, land use, or settlement pattern changes in a period of a few years or decades. Older aerial photographs, topographic maps, and other archival sources provide information about an area at a given point in time. Comparing this with present information provides data as to the nature and rate of change. If maps or aerial photographs are available at five- or ten-year intervals, the scope and speed of change can be documented with a high degree of accuracy. The use of current as well as historical Sanborn cadastral maps is a useful data source for urban change documentation. As in a time lapse view of the opening of a flower, the sequent occupancy of a place may be identified and measured. For many studies concerned with historical geography, comparing of secondary sources of information with primary sources obtained through field mapping and measurement is a viable and well-established method of research.

## Observation and Verbal Commentary

In many situations, the most efficient way of documenting observable landscape features is through verbalization. The tape recorder is, for many field workers, replacing laboriously written field notes. While clearly an advantage in time-constrained situations (such as during aerial reconnaissance and regional parcel mapping, when motion and rapidity of movement preclude external written notes), recording vocal comments is useful in all phases of field work. Rapid dictation leads to more comprehensive description and greater cost effectiveness for each hour spent in the field. Verbal notes can be transcribed easily in the laboratory or office at greatly reduced costs. In addition, the tape recorder is always available and easily transported. With it, peripheral, but valuable observations not necessarily planned for in the research design may be noted. The various tape recordings made by students and assistants insure that the information reaches the principal researchers. They may also give a unique perspective to those persons not able to participate in the data gathering.

Tape recordings also permit the researcher to return to the field vicariously so she can refresh her memory or refine descriptions of field phenomena. This is valuable especially when a long interval elapses between data gathering and the analysis and synthesis of field research projects.

## Mapping Technique Selection

The mapping techniques described previously are to be considered only as representative examples. The researcher should feel free to modify these techniques or, with a dash of creativity, devise new approaches to suit a specific field problem. However, the development of an innovative technique is not the goal in itself, but the means to collect new data accurately and efficiently, which is, in the last analysis, the basic purpose of all field mapping.

## SUGGESTED REFERENCES

AVERY, T.E., *Interpretation of Aerial Photographs.* Minneapolis, Minnesota: Burgess Publishing Co., 1968.

CASE, R. P., "Microcomputers in the Field," *Byte*, June 1985, 243–250.

CHOW, V. T., *Handbook of Applied Hydrology.* New York: McGraw-Hill Book Co., 1964.

CLARK, A. H., "Field Research in Historical Geography," *Professional Geographer,* Vol. 4, 13–23. Washington, D.C.: Association of American Geographers, 1946.

COLLINS, M. P., "Field Work in Urban Areas," *Frontiers in Geographical Teaching,* edited by R. J. Chorley and P. Haggett, Chapter 10. London: Methuen Publications, 1965.

FINCH, J. K., *Topographic Maps and Sketch Mapping.* New York: John Wiley & Sons, 1920.

GREENHOOD, D., *Mapping.* Chicago: University of Chicago Press, 1964.

GUNN, A. M., *Techniques in Field Geography.* Toronto, Ontario: Copp Clark Publishing Co., 1962.

*HANDBOOK OF METEOROLOGICAL INSTRUMENTS.* Air Ministry Meteorological Office. London: Her Majesty's Stationery Office, 1956.

HARING, L. L. and J. F. LOUNSBURY, *Introduction to Scientific Geographic Research.* Dubuque, Iowa: William C. Brown Co., 1975.

HEDGE, A. M. and A. A. KLINGEBEIL, "The Use of Soil Maps," *Soil: The Yearbook of Agriculture, 1957,* 400–411. Washington, D.C.: U.S. Government Printing Office, 1957.

HOFFMAN, G., "Hints on Measurement Techniques Used in Microclimatological and Micrometeorological Investigations," *The Climate Near the Ground,* by R. Geiger. Cambridge, Massachusetts: Harvard University Press, 1965.

KUCHLER, A. W., *Vegetation Mapping.* New York: Ronald Press Co., 1967.

LOW, J. W., *Geologic Field Methods.* New York: Harper & Row, Publishers, 1957.

————, *Plane Table Mapping.* New York: Harper & Row, Publishers, 1952.

MURPHY, R. E., *The American City: An Urban Geography.* New York: McGraw-Hill Book Co., 1966.

PLATT, R. S., *Field Study in American Geography: The Development of Theory and Method Exemplified by Selections,* Department of Geography, Research Paper No. 61. Chicago: University of Chicago Press, 1959.

PUGH, J. C., *Surveying for Field Scientists.* Pittsburgh, Pennsylvania: University of Pittsburgh Press, 1975.

SHIMWELL, D. W., *The Description and Classification of Vegetation.* Seattle, Washington: University of Washington Press, 1971.

SMITH, G. D. and A. R. AANDAHL, "Soil Classification and Surveys," *Soil: The Yearbook of Agriculture, 1957,* 396–400. Washington, D.C.: U.S. Government Printing Office, 1957.

# Acquisition of Nonvisible Information

## Chapter

# 4

# Chapter 4

In many geographic field problems, the research geographer must gather data that are not part of the visible landscape. For example, in an agricultural study, the reseacher must acquire information about the rotation of crops, farming practices, marketing locations, etc., to develop a complete understanding of the agricultural economy. Perceptions, attitudes, and behavior patterns are other forms of nonvisible data that may be critical to a given field problem. Mapping the visible features of the landscape can be done with a high degree of accuracy, assuming that the classification systems, scale, and base maps are compatible. Acquiring nonvisible information, however, is often less precise. Interviewing methods must be employed to obtain this kind of data, and misunderstandings may occur in numerous ways. The respondent, that is, the person being questioned, may not fully comprehend the question being asked, or the interviewer may misinterpret or misunderstand the response. The respondent may not have an answer to the question, but will answer nevertheless; or, for one or more reasons, will purposely provide a misleading or deceiving response; or will answer without giving the question much thought.

The science (and art) of interviewing is a complex and highly specialized professional field in itself. In sophisticated studies, the research geographer should consult with specialists in the field for advice in developing an interview survey design. The accuracy of the data obtained by interviewing is largely dependent on:

1. The design of the questionnaire or the questions to be asked informally;
2. The manner in which the questionnaire or questions are administered;
3. The design of the sampling procedure, if everyone in the research area is not to be surveyed; and
4. The method by which the responses are compiled and tabulated.

## QUESTIONNAIRE DESIGN

The basic objectives in the design of a questionnaire are twofold:

1. To obtain information relevant to the research study undertaken, and
2. To acquire this information with the highest degree of accuracy possible.

Whether the information is to be obtained by talking with people informally or by constructing a questionnaire, the information desired must be precisely defined. Exactly what data are essential to the specific research problem and what are not must be identified as clearly and specifically as possible.

### Open Versus Closed Responses

The nature and scope of a given problem will determine how, and to what degree, the answers or responses to the questions asked should be classified. The open response and the closed response forms of the questionnaire represent the two extremes. The *open* or *free response form* allows complete freedom in answering (see Figure 4-1), and the answers are not classified in a precise manner. It is most helpful in reconnaissance and pilot studies because it reveals the range of possible answers that could be given to a particular question and provides certain qualitative information which could be used to develop a closed response questionnaire for subsequent interviews. Furthermore, the responses obtained may serve as "quotable" material to add "color" or the "feel of the area" to the research report. However, the spontaneity, selectivity, length, and general lack of organization of the responses make coding and precise statistical analysis difficult.

A variation of the open response form which is useful in certain geographical field studies might be called the *discussion interview*. This type of interview is a completely open discussion. The researcher might direct, but not control, the conversation, and specific questions are not formulated in advance. In geographic field work, it is highly desirable that the researcher conduct the open response and discussion forms of interviews. Closed response forms of questionnaires may be administered by a properly trained interviewer.

The closed response form limits the responses to specific answers, which are classified in detail and can be analyzed statistically without undue difficulty. Closed response forms have other advantages over open response forms. They normally are shorter and require less time to administer, they require less skill and effort on the part of the interviewer, and they make it easier for the interviewer and respondent to keep on track. The disadvantage of using the classified response

OPEN RESPONSE (open ended, no classification)

There has been some concern about storm damage in this area in recent years. During the last two years, has your house or property been damaged?

_____
_____
_____
_____

CLOSED RESPONSE (classified response)

There has been some concern about storm damage in this area in recent years. During the last two years, has your house or property been damaged in any of the following ways?

[ ]   Basement flooded.
[ ]   Trees uprooted.
[ ]   Windows broken.
[ ]   Roof damaged.
[ ]   Yard flooded.
[ ]   Other (specify): _____

### Figure 4-1.

Examples of Open Response and Closed Response Questions.

format is that it may produce answers when none exist, such as when the respondent has no information or experience in the area under study, or in the case of attitude surveys, when he or she has no opinion.

The researcher, in exercising some creativity, can develop a questionnaire that combines the advantages of both the open and closed response forms by using unstructured and general questions along with structured questions. The use of either an open response, closed response, or some combination thereof, is dependent on the nature and magnitude of the specific problem.

### Question Writing

The objective of question writing is the same as that of questionnaire design as a whole: to acquire complete, accurate data that are pertinent to the study undertaken. The perfect question will invoke the cooperation of the respondent, show respect for his or her privacy and dignity, and at the same time produce the information desired. Careful thought must be given to the exact wording of the questions to avoid misunderstandings, which result in unreliable information. The following generalized checklist may help you avoid the common pitfalls:

1. *The question should be stated as clearly, concisely, and specifically as possible.* Extremes in vocabulary should be avoided, and words should be selected that are direct, simple, and familiar to both the interviewer and the anticipated

respondents. Slang, colloquialisms, jargon, and technical terms should not be used. The wording and phrasing of the questions should be understandable to people in all walks of life and at all educational levels. The questions should not be too complex or too general, and ambiguity should be avoided at all costs. Indefinite words and phrases such as "usually," "fairly near," "on occasion," lack precision and mean different things to different people, and therefore should be avoided. The researcher must also bear in mind that the usage and meanings of common words differ from one region to another. Questions also should be carefully scrutinized and excessive wording eliminated. Short, direct questions of twenty words or fewer are the most effective. Conditional clauses and qualifications should precede the key idea so that the core of the question should come last.

2. *The question should not be leading, loaded, or double-barreled unless there is a purpose.* A leading question directs the respondent, either by suggestion or implication, toward a specific answer. Such phrases as "as a knowledgeable person, don't you agree" or "wouldn't you say" push the respondent in the direction of a certain response. A loaded question directs the answer by using stereotypes, emotionally charged words or phrases, prestige suggestions, etc. Also, questions may be loaded by mentioning the desired alternative such as, "What do you do for recreation?—picnic or what?" Double-barreled questions contain two or more issues. For example, "Do you plan to give up farming and look for other work in the near future?" A "yes" could mean that the respondent does plan to leave farming and look for other employment, or plans to give up farming and not look for other work, or plans to look for other employment but not give up farming. And the respondent's perception of the "near future" is anyone's guess. Loaded and double-barreled questions have distinct disadvantages and will result in imprecise data. However, leading questions may increase the validity of certain research, particularly in studies dealing with behavioral patterns. In these cases, the researcher must first recognize that a question is truly leading and then decide whether it will benefit or detract from the overall survey.

3. *The question must be applicable to all respondents.* Inapplicable questions such as, "What do your children do for recreation?", when the respondent has none, are not only irritating but also may be misleading. The respondent may give an incorrect answer to oblige the interviewer. An incorrect answer may also be given when the question is such that the true answer might be injurious to the respondent's self-esteem. The standard solution to this problem is to include a *skip pattern* in the questionnaire. For example, the first question might be a qualifying one such as, "Do you have any children living with you?" If the answer of a specific respondent is "no," and questions 2 through 5 are concerned with the behavior of children, the interviewer would skip, or ask the respondent to skip, to question 6, which might be applicable. Respondents who have children would be asked questions 2 through 5. In complex research studies, the use of skip patterns, filters, or pivot questions can be quite complicated. However, a soundly designed set of skip patterns will eliminate the need for several sets of questionnaires for different types of respondents, minimize confusion in the field, and facilitate the analysis later.

4. *The response style employed must be compatible with the intent and content of the question.* For certain questions, a simple "yes-no" or "agree-disagree" response format will suffice. Usually, however, a simple "yes" or "no" will not provide the desired data. In these cases, other styles can be employed. The

common styles, or sets, are: (1) *checklist types*, where a number of possible answers are given and the respondent checks all that apply; (2) *intensity scales*, where the respondent is asked to rate an event on a single dimension of intensity or quantity, ranging from more to less; (3) *frequency scales*, where the respondent rates an activity or event in terms of time factors; (4) *distance scales*, where the respondent rates distance from one point to others; and (5) *ranking scales*, where the respondent arranges a series of options in order of personal preference or some other standard (see Figure 4-2).

---

CHECKLIST TYPE (variety of possible answers; all those that apply are checked)

During the last five years, has the production of soybeans on your farm decreased because of too little rainfall?

[ ] No noticeable decrease during this period.
[ ] Slight decrease some years.
[ ] Slight decrease every year.
[ ] Serious decrease some years.
[ ] Serious decrease every year.

---

INTENSITY SCALE (quantity, intensity, "how much")

Airport noise is a serious health hazard in this area.

[ ] Agree strongly with this statement.
[ ] Agree moderately with this statement.
[ ] Undecided.
[ ] Disagree moderately with this statement.
[ ] Disagree strongly with this statement.

---

FREQUENCY SCALE (frequency of an activity, "how often")

During the last year, how often did you visit Black Lake Recreational Park?

[ ] At least once a week.
[ ] Once a month.
[ ] Once every two or three months.
[ ] Once or twice a year.
[ ] Not at all.

---

DISTANCE SCALE (distance from one activity to another, "how far")

How far do you travel (one-way-trip) to work?

[ ] Less than one mile.
[ ] One to 4.9 miles.
[ ] 5 to 9.9 miles.
[ ] 10 to 24.9 miles.
[ ] 25 miles or over.

---

RANKING SCALE (priority, "relative importance")

The opening of the Metro Medical Center has had some impact on this area. Which of the following do you consider to be most serious ("put "1" in the space next to that item)

[ ] Traffic congestion.
[ ] Noise.
[ ] Shortage of parking areas.
[ ] Lowering of property values.
[ ] Unsightly structures; detracts from the beauty of the neighborhood.

If you could give *two* answers, which of the items above would you choose second? Put "2" next to that item.

If you could give *three* answers, which of the items above would you choose third? Put "3" next to that item.

---

Figure 4-2.

Examples of Response Styles.

A great number of scales can be devised. The researcher must be careful to select or devise a scale that is compatible to the question asked and that is easily understood and definable by both the interviewer and the respondent. There is some evidence that the answer to a given question could be influenced by the type of response styles employed. However, no generalizations can be made about such influences. The researcher should be aware of this possibility and, as far as it is possible, design response styles that will not affect responses.

## Question Sequence

The researcher also should consider the sequence of questions. Logic and common sense will determine what questions come first and what questions follow in most cases. The first questions may be of a general nature, easy to respond to, and of the type to excite the respondent's interest and motivation to participate. One approach is to use questions of a broad and general nature—perhaps unstructured questions— and proceed to structured questions that are more specific. The merit of this *funnel sequence* of topics and questions is that it allows for a freedom in response in the beginning but later arrives at questions that will provide structural data. In some studies, an *inverted funnel sequence* is effective, where specific and structured questions are asked first and are then followed by questions concerning the broader or more general issues. Such an approach may be employed when the respondent is not likely to have a formulated opinion on the broader issue or when the respondent has such strong feelings on the general topic that the subsequent classified responses to specific questions would be biased.

## Sensitive Questions

When designing a field research problem the researcher must sometimes consider unpleasant or sensitive topics. If at all possible, sensitive issues should be avoided; but if such issues are vital to the study, they should be handled with extreme care. Invading a person's privacy is always an ethical and, in some cases, a legal transgression.

No foolproof guidelines can determine whether a given question is sensitive or not. Obviously, some questions would be embarrassing or unpleasant to all potential respondents. Other questions may be sensitive to some respondents and not to others. Sensitivity will vary from place to place, from time to time, from age group to age group, and in some cases, from one sex to the other.

If the researcher must ask sensitive questions, it is best not to place them at the beginning of the interview. It is important to create a feeling of trust and good will between the respondent and interviewer as soon as possible, and only questions that will lead to mutual confidence and trust should be asked. On the other hand, touchy or prickly items should not be placed at the end of the questionnaire or interview either. It is important to leave the respondent in a positive and happy frame of mind. Therefore, sensitive questions should be placed in the body of the questionnaire only after a feeling of trust and rapport has been developed. The appropriate point to ask these questions might vary from one respondent to another or from one interviewer to the next. If the interviewer can sense the right time, and the questionnaire has some flexibility, the right time might be after four or five questions or perhaps not until twenty or more items have been completed. Furthermore, it is desirable to bring up sensitive questions in the section of the questionnaire where they seem reasonable in relation to immediately preceding queries. Approaching the problem questions gradually with warm-up items is often useful.

Great care should be exercised in dealing with unpleasant topics; it is best to eliminate them entirely if doing so does not jeopardize the overall study. If they cannot be deleted because of their value to the field research, they should be worded in the least offensive manner possible.

## Skip-Structured Questionnaires

In some surveys, every question may be applicable to all respondents. Designing a skip-structured questionnaire can speed up the interview and increase the efficiency of the survey. For example, if a field study is concerned with the nature and amount of property damage due to recent storms in a local area, respondents may be divided into two groups (see Figure 4-3). The key question that separates the respondent population is sometimes referred to as the sieve question. Skip-structured designs solve the problems of asking questions that do not apply to a given group of respondents.

## Numbering and Coding

All questions or items should be numbered consecutively throughout the questionnaire to expedite its administration in the field and to facilitate the analysis of responses. Furthermore, each questionnaire should have space to record identifying numbers, location of the specific interview, date of administration, and name or code number of the interviewer if more than one is involved in the research study. The interviewer should mark each questionnaire with an identifying number as it is given and also record this number on the base map. Geographic field work emphasizes the spatial distribution of phenomena, and where data are acquired is most critical. The exact method of coding will depend on whether or not the tabulations are to be done by hand or computer, and if computerized, on the types of software and computer facilities that will be used.

## ADMINISTRATION OF THE QUESTIONNAIRE

There are four major methods of administering questionnaires or giving interviews. These are:

1. The telephone interview;
2. The mail questionnaire;
3. The group self-administered questionnaire; and
4. The personal interview, in which the respondent and interviewer are face-to-face and in a one-to-one situation.

Each method has its advantages and limitations, and, as is true with many other aspects of geographic field work, the most effective method or combination

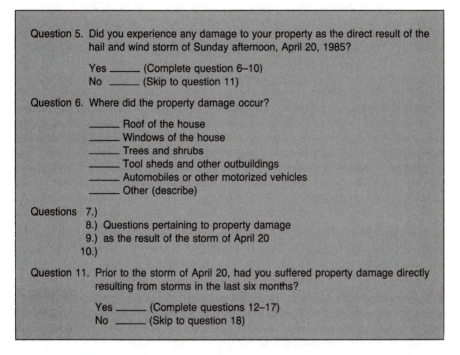

Question 5.  Did you experience any damage to your property as the direct result of the hail and wind storm of Sunday afternoon, April 20, 1985?

     Yes _____ (Complete question 6–10)
     No  _____ (Skip to question 11)

Question 6.  Where did the property damage occur?

     _____ Roof of the house
     _____ Windows of the house
     _____ Trees and shrubs
     _____ Tool sheds and other outbuildings
     _____ Automobiles or other motorized vehicles
     _____ Other (describe)

Questions  7.)
        8.) Questions pertaining to property damage
        9.) as the result of the storm of April 20
     10.)

Question 11. Prior to the storm of April 20, had you suffered property damage directly resulting from storms in the last six months?

     Yes _____ (Complete questions 12–17)
     No  _____ (Skip to question 18)

**Figure 4-3.**

An Illustration of a Skip-Structured Questionnaire.

of methods is determined by the nature and magnitude of the specific problem.

The *telephone interview* is relatively inexpensive and highly flexible. It can be used to advantage, particularly in cases where the questionnaire is short and direct and does not deal with sensitive or personal issues. When used to obtain supplementary data in following up a person-to-person interview, it may be most effective. In these cases. the precise location of the respondent's place of residence has been determined and initial contacts have been made previously. Conversely, the telephone interview may be employed to contact and prepare the prospective respondent prior to a personal interview. Finally, the telephone interview may be used to fill in areal gaps and to further refine the spatial patterns established by other methods.

The major disadvantage of the *telephone interview* in geographic field work is that it is always difficult, and sometimes impossible, to match a telephone number to a location that can be accurately recorded on a base map. Consulting a telephone directory for an address is time-consuming and may not resolve the problem. In rural areas, particularly, addresses may be given as post office boxes or rural route numbers. In addition, there has been a widespread increase in recent years in the use of telephones to conduct marketing surveys, some associated with promotion schemes, advertising campaigns, and sales presentations. Potential respondents may be highly suspicious and reluctant to cooperate. It also is difficult to establish the same rapport on the telephone that is possible in a person-to-person situation.

The major advantages of the *mail questionnaire* are its low cost and ease of distribution. It is perhaps for these reasons that this format has been greatly overused in recent years, causing the public to become complacent. Returned responses of 30 to 40 percent are considered normal, while a response rate of 70 percent or more is exceptional. Furthermore, the respondent has complete freedom whether or not to respond, and as a result, the returned responses may represent only those individuals who have strong feelings one way or another. Obtaining a true sample or cross section of the population surveyed, therefore, is unlikely, particularly if the overall response rate is low. The response rate to a given questionnaire is directly related to the respondents' interest in the research subject, the length and difficulty of the questionnaire, and the amount and success of advance publicity or notice in preparing individual respondents and the community as a whole.

Several mechanisms have been used, with varying degrees of success, to increase response. The most effective include (1) providing an addressed and stamped envelope for returning the completed questionnaire, (2) mailing the questionnaires by first class mail, (3) including a personally typed cover letter to describe the need and value of the study in order to increase the interest of the respondent, (4) reproducing the questionnaire on colored paper rather than white, and (5) using follow-up phone calls and/or postcards. The mail questionnaire should be as short and direct as possible, request only information not available from other sources, and ask only important questions relevant to the study.

Although it requires more effort, a highly effective method for some types of studies is for the interviewer to visit the respondent and explain the nature of the study and request that the respondent return the completed questionnaire by mail in a provided addressed, stamped envelope. This method is essentially a combination of the mail and self-administrated types. It has the advantage of face-to-face contact as well as of letting the respondent choose the time and place to work on the questionnaire. As a result, the questions and responses likely will be given more thought.

The *group self-administered questionnaire* has the distinct advantage that the interviewer is present to provide motivational impetus and moral support. The self-administered questionnaire is normally completed in group sessions in a conference room, office, or classroom. Two of the limitations inherent in this method are that the respondents can, and do, look ahead, disrupting any orderly sequence of questions that might exist, and that they can compare their responses with others. Further, in most cases, people that have gathered together have done so for reasons other than to complete a questionnaire. Normally a group of respondents might represent a select group—students, members of a certain organization, and so forth. For this reason, the group self-administered method is limited to studies that do not require a cross section of the population.

Although the most costly method in time and dollars, the *personal interview* is the most effective means to obtain information, both quantitatively and qualitatively. The person-to-person situation creates a setting in which ambiguous answers can be clarified. It lets the interviewer control the question sequence and probe for additional details to any or all of the questions to improve the quality of the data. Further, the personal interview is not as dependent on the educational and literacy level of the respondents as other methods.

The flow of information in the personal interview depends upon: (1) the respondent's interest in the subject of the research study; (2) the immediate environment in which the interview is being conducted; (3) the skill, personality, and motivation of the interview; and (4) the willingness and ability of the respondent to participate. There are no two identical interview situations. There is no perfect interviewer for each and every occasion; no uniform or standard respondent; and the subject usually varies from one research study to another, or if not, the values and attitudes of the community and respondents vary from place to place and from time to time. The interrelated factors which either facilitate or obstruct the flow of information from the respondent to the researcher are illustrated schematically in Figure 4-4. The ideal conditions include: a reseach topic that does not concern itself with sensitive or emotional issues and is of more than passing interest to the respondent; an interviewer with a pleasing personality who is a sympathetic, highly motivated, and capable individual; a respondent that is most willing to cooperate and has the ability and desire to provide relevant, precise, complete, and honest answers; and an interview that is conducted in a harmonious atmosphere free from distractions. Rarely will such conditions exist in field situations.

## Discussion Interview

A completely open response type of interview is of great value in some geographic field problems. In this *discussion interview*, the respondent is encouraged to discuss opinions, insights, facts as he or she recalls them, etc., with little or no restrictions. Such an interview might be described as a general conversation with an ulterior motive. The researcher *is* seeking to acquire relevant information but does not attempt to control the discussion, or classify or structure the responses. In the discussion interview, the researcher plays a role similar to a reporter interviewing a celebrity passing through town. The information may not be totally accurate in all cases, but it may provide data relevant to a specific research problem that cannot be acquired in any other way.

This method may provide valuable data in at least two types of field situations. In the first instance, the field researcher is seeking general information concerning the research area in order to get a "feel" for the region, enrich his or her background, or develop an understanding of the locality. The discussion interview is useful when the researcher is attempting to acquire an overall perspective of the area to supplement a specialized and specific study. For example, if the research problem is to analyze the microclimates in a given area, the researcher may wish to learn about the impact of climate on the agricultural economy and other activities within the area in general rather than in specific terms. Or the researcher may not have a great deal of information about the area and is accumulating qualitative data to be used as the basis for developing a structured design for subsequent research.

In the second instance, the discussion interview is directed to selected individuals who possess information that is not known to others. For example, the researcher concerned with a problem in historical geography may increase his or

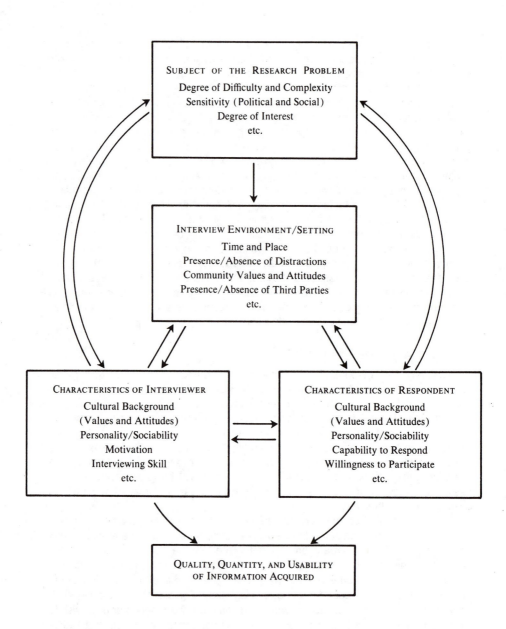

**Figure 4-4.**

Interviewer-Respondent Interrelations and Factors Affecting the Communications Process.

(After D. P. Warwick and C. A. Lininger, *The Sample Survey: Theory and Practice,* McGraw-Hill Book Company, 1975.)

her knowledge about the problem by talking with the old, or perhaps even original, settlers in the area. Such persons may have information and insights about the early history and settlement of the area that cannot be obtained from other sources. Additional examples might include talking with hunters or trappers about the intensity, movements, range, etc., of wildlife; or discussing the impact on an area of a proposed new industrial plant with a governmental official or influential citizen. Most frequently, the numbers of these selected people in a given area are few and, in total, do not constitute a large enough population to justify having their comments statistically analyzed.

The researcher must be sufficiently informed about the subject to be able to discuss the topic intelligently. He or she should determine the kinds of information desired, perferably in a logical sequence. On the other hand, the researcher must be able to formulate on the spur of the moment questions that may reveal unforeseen but pertinent information. In most cases, such questions cannot be framed in advance.

Generally the researcher should make an appointment with the respondent prior to the actual interview. This will provide the respondent time to collect his or her thoughts and perhaps prepare in advance of the discussion. In other situations, the person to be interviewed may be subject to a tight and busy schedule and must plan in advance.

The discussion interview is most effective when the researcher does not attempt to structure or control the discussion but, at the same time, keeps the conversation from moving too far afield. The researcher may jot down a few notes during the interview if it is not disruptive, but as soon as possible should make as complete a record as possible of the conversation. The use of a tape recorder is invaluable in these types of interviews. It is highly effective in studies where the objective is to explore, in depth, complex issues with a limited number of respondents. The transcribed responses provide a verbatim record of the conversation and eliminate the potential problems of selective recall and distortion. Use of a tape recorder also leaves the researcher free to concentrate on the discussion of the moment and provides a mechanism whereby the entire conversation can be studied at a later time. The tape recorder should be as inconspicuous as possible to avoid distractions.

Another form of gathering data related to the discussion interview is sometimes called experiential field work. This technique has been used to gather spatial data about the life experiences of the elderly.*

In normal interview techniques, the researcher makes a major effort to maintain objectively, to refrain from using leading questions, and to ensure that he remains as intellectually remote as possible from the respondent. The experiential technique is completely different. The researcher enters into the experiences of the respondent through intense personal interactions as a friend and confidant. More of an "attitude" than a technique, it requires both parties to come to a mutual "interpersonal knowing" by trusting and caring for each other and sharing emotional experiences.

---

*An excellent reference describing this technique is G. D. Rowles' "Reflections on Experiential Field Work," Chapter 11, *Humanistic Geography: Prospects and Problems* (Chicago: Maanoufa Press, 1978).

The idea is to select a few individuals and develop a close personal friendship with each. Over an extended period (possibly years) the researcher makes tape recordings and notes of all conversations. Both individuals attempt to explore the subject matter of interest in the research and to develop a mutual language for communicating, describing, and perceiving geographical dimensions and cognitive involvement.

Some drawbacks of doing experiential field work include the intense personal involvement with the respondents, the length of time that must be invested in a project (possibly years), and the small number of respondents.

The researcher becomes a translator between subjective experiences, attitudes, and the lexicon of spatial interaction. Thus, inductive inference from tape recorded conversation becomes the essence of the procedure. Subjective inquiry leaves the informant and the researcher changed by the experience in emotional and intellectual ways. For research involving cognitive and spatial perceptions, spatial attitudes, and viewpoints, this approach may result in data which is unique.

### Pilot Tests

Pilot or pretests are advisable for several reasons. Field testing or trial runs usually reveal problems with the testing instrument itself or the manner in which the questionnaire is administered. Fifteen to twenty-five interviews representing a good cross-section of the total respondent population as well as a good areal distribution will normally suffice to isolate or define problems in the overall survey. Pretests may reveal

1. the likelihood of a controversy arising from the survey itself,
2. the number of refusals and the reasons for a participant's refusal,
3. the estimated cost and effort to conduct the study,
4. whether or not the wording of questions is clear and unambiguous,
5. if there are local differences in the interpretation of questions (This may be a problem, particularly if the research area is large and includes a variety of social, ethnic, language, or religious groups.),
6. which items, if any, are difficult for the respondents to answer,
7. the reasons why some questions are sensitive or unpleasant,
8. the effectiveness of skips and filters,
9. whether or not the sequence of questions is logical, smooth, and workable, and
10. the effectiveness of the proposed sampling system.

Pilot or pretests are necessary proving grounds for the overall field study; they are well worth the extra time and effort. Obviously, the more thought and effort put into developing the questionnaire and survey, the less likelihood of encountering serious problems during the pretest period.

### Validity and Reliability of Interview Data

Highly sophisticated mechanisms for determining the measurement adequacy of interview data are normally not necessary in geographic field studies and are beyond the scope of this text. If such data are necessary, the novice researcher should consult with specialists in this field. The important aspects of measuring the adequacy of a given questionnaire and interviewing procedure are *validity* and *reliability*. A third aspect is *precision*. Validity is the degree to which a questionnaire does in fact measure what it purports to measure. For example, if interviewing data concerning voting behavior obtained prior to an election corresponds closely to actual election results, the interviewing instrument provided the data for which it was designed and may be considered valid. The reliability of an interview instrument refers to whether or not the same information is obtained repeatedly in identical circumstances. A reliable instrument will yield the same results consistently if conditions remain constant. Precision is the ability of a given questionnaire to reveal small differences in opinions and attitudes. Each questionnaire has its limitations beyond which it is not possible to detect shades of meaning of responses. The specific study will determine how sensitive the interview question need be. The relationships between validity, reliability, and precision are complex, and each aspect is neither wholly independent of nor simply interrelated to the others.

One measure of reliability is to code the interview data previously obtained and then repeat or retest a given question or item under identical circumstances, or, if numerous items are involved, compare one-half of the items, randomly selected, with the results obtained by using the other half of the items (split-half coefficients). Validity may be measured by comparing the data collected with some other existing standard measure. If the two measures agree, we may assume that the validity of the item tested is at least as valid as that of the measure used as the standard. For example, interviewing data on savings accounts ought to agree with bank records. The validity of factual data is much easier to measure than data concerning personalities and attitudes. In short, the careful design of questions and response styles and the application of the conditions necessary for a successful interview previously discussed are perhaps the best overall mechanisms for achieving validity and reliability of interview data.

## Preparing the Respondents and Community

In geographic field studies, advance contacts in the area where the research will be conducted are helpful. This applies to all studies using mapping only, interviewing only, or a combination. The researcher's presence may give rise to a great deal of speculation among the local residents, particularly in rural areas. At least the local police or sheriff, Better Business Bureau, and/or Chamber of Commerce should be informed as to the locations of the proposed research, the approximate dates of the field work, and the general nature of the research study. This is an important aspect of field work today since an increasing number of communities recently have approved ordinances that control house-to-house solicitation. In some cases, these ordinances include survey research.

In the smaller cities and towns, it may be useful to contact appropriate local government officials and service organizations. At the very least, such contacts are good public relations and might lead to active support. In most communities, the local service organizations are composed of civic-minded and influential citizens whose attitudes could facilitate or obstruct research work. However, the major purpose of advance contacts is not to obtain approval for the research study as such, or to seek official support, but rather to serve as a mechanism of information to notify the people and groups in the area of the research project and to solicit their cooperation.

It is very helpful to have the local newspapers print a story describing the research study. The ideal time would be a few days in advance of the study; a report too far in advance or after the actual field research is underway may be ineffective. In general, newspaper publicity is most helpful when it is "low profile." Opposition and resistance to the study could develop as a result of an overly organized and large-scale publicity campaign.

The field researcher should bear in mind that some individuals are suspicious of investigative studies of any type. They may think that the field researcher is actually an investigator for some agency of the government or insurance company, a person "casing" the property in preparation for a burglary, or a representative of an unscrupulous business enterprise. This is an understandable attitude in view of the rapid increase in crime, fly-by-night business operations, fraudulent land and development schemes, etc. The researcher, insofar as possible, should allay these suspicions and stress the legitimacy of the research study.

Furthermore, the researcher should take the viewpoint that he or she is the recipient, in the sense of taking information from the area for personal benefit; and that the respondents are the donors, in the sense that they are giving information, time, and effort to further the researcher's cause without deriving any tangible gains. Common courtesy, friendliness, and an understanding attitude expedite the research work in the great majority of cases.


## SAMPLING

All geographic field research problems are sample studies in the sense that it is not possible to obtain information for the entire spectrum of a given area, regardless how small. Only certain phenomena are selected for study from a range of hundreds of possibilities. Further, the selected phenomena that exist specifically in the field are classified into generalized groupings, and the data are collected at a scale considerably less detailed than reality.

In addition, any given research area is a portion of a larger area, and even if all the data in a research area were obtained, they would represent only a sample of a larger universe. Field research, like most other research, consists of sample observations from which understandings or perceptions of the total situation are inferred. How close the inferences are to real conditions depends on the kind and number of observations and the manner in which they were made.

When using interviewing procedures to collect information, it is most desirable to obtain total coverage of the research area. However, in most field studies, the total population is so large that not all can be interviewed due to time and cost restraints. In these cases, a sample procedure must be devised. A sample is a percentage of the total, and the problem facing the researcher is to select the right percentage, both in kind and number, that will truly represent the total picture. There are no standard sampling designs that are effective for each and every problem. A sampling procedure should be devised to be compatible to the objectives of the specific problem and the kinds of information to be collected. It need not be highly complicated, but it must be systematic and orderly. The researcher must bear in mind that the primary objective of geographic field work is to collect data pertinent to the problem and not to create a complex and theoretical sample design. Sampling in itself is not the goal; it is the means whereby the primary objectives of the research may be accomplished.

All geographic field work is spatially oriented. Therefore, any system of sampling must be spatial in design. All parts of the research area must be considered, although not necessarily equally, in the development of a sampling design. For example, if the researcher were attempting to determine the place of work of the nonfarming population in a rural area, he or she might structure a sampling procedure whereby one interview would be given for each five nonfarm rural dwellings. The rural nonfarm dwellings could be grouped into units, each encompassing five dwellings and, by using a table of random numbers, the specific dwelling in each unit could be determined. Such a sampling procedure would be spatial in that areal differentiations would be revealed and all of the research area had been considered. It is not important that the same number of interviews be given in each square mile. The density of rural nonfarm dwellings might be greater in some parts of the area than others; perhaps large portions of the research area have no nonfarm residents at all, and for these areas no interviews could be given. However, this example of a simple sampling procedure is systematic, and the entire research area has been covered. The intensity of interviews varied from one part of the area to another, reflecting the existing settlement pattern on which it was based.

This hypothetical example is concerned with only the place of residence versus the place of work. If the problem was to determine whether or not the place of work could be correlated with the economic or social level of the nonfarming gainfully employed population, the sample would need to be stratified. The nonfarming population would have to be classified as to income groups, level of education, etc., and each group sampled in order to obtain a true cross section. Opinion polls and marketing surveys are examples of types of stratified samples. The accuracy of the data obtained is directly related to the manner in which the groups were classified as well as the percentage of persons surveyed within each group.

Sampling procedures pertinent to geographic field studies will be discussed in the next chapter. In all cases, however, the researcher must consider the following in developing a sampling system for interviewing:

1. The type of data desired;
2. The degree of accuracy the researcher wishes to attain;
3. The importance of an economic, social, or cultural cross section of the population to be studied; and
4. The way in which the data are to be analyzed and applied.

## COMPILATION AND TABULATION

The task of compiling the responses derived from a closed response form of questionnaire may be time-consuming but not difficult. The numbers and percentages of each response for each question can be tabulated by hand or computerized. In the latter case, the questionnaire must be properly coded to be compatible to the software and computer facilities being used. Compilation of the open response form of questionnaire is a great deal more difficult. This form of interviewing produces a surprisingly large variety of responses, both in terms of content and length. To classify these diverse responses into a form that can be statistically analyzed, the researcher must separate the relevant from the irrelevant in some systematic manner. For example, if the problem was to survey opinions regarding the impacts of a proposed new land-use plan, the researcher might develop a classification of responses such as "strongly support," "support with reservations," "uncertain," "strongly disapprove," etc. The researcher may go further and classify the reservations or specific items of concern into groups such as "lowering property values," "increased traffic congestion," etc. Having developed this structure, the researcher must carefully analyze each response to separate and tabulate the components according to the classification system established. The obvious problem facing the researcher in compiling and tabulating open questions lies in the development of a system which both encompasses the wide range of answers and provides categories for statistical analysis.

The compilation of interviewing data in geographic field work must include matching a set of data with locations on a map. Each interview given must be precisely located in order that spatial patterns can be revealed in subsequent analysis. One method is to construct a grid pattern on the base map (a transparent overlay would suffice) and code each interview areally. The areal or spatial aspects of geographic field work are what differentiates it from other types of field research. For example, a hypothetical and very simple problem might be concerned with perceptions of noise generated by airports in a research matrix of 12 to 15 miles in radius. The question might be, "Do you regard noise as a serious health hazard in this area?" with the responses limited to "yes," "uncertain," and "no." The tabulation of the responses might reveal that there is no dominant opinion for the research area as a whole (see Figure 4-5A). In matching the data to locations, a spatial pattern might be revealed (see Figure 4-5B). The "yes" and "uncertain" responses decrease with distance from the airport, while the "no" responses increase.

The spatial pattern can be further refined by presenting the data cartographically rather than in a statistical table form (see Figure 4-5C). It is now revealed that the opinions regarding airport noise are not spherical in areal extent. The percentages of "yes" responses extend eastward considerably further than other directions. The explanation might be associated with flight patterns; but whatever the causes, the researcher then must account for this areal distribution to fully complete the research problem.

## Selection of the Interviewing Procedure

The acquisition of information not readily observable or apparent presents several problems for the researcher. The first critical and basic decision is to determine precisely what data are essential, and what are not, to the solution of a given problem. This decision bears directly on subsequent questions concerning the form

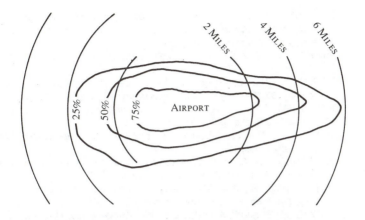

| A. TOTAL RESEARCH AREA. | | B.PERCEPTION/DISTANCE RELATIONSHIPS. | | | | |
|---|---|---|---|---|---|---|
| | % of Total Responses | | Within 2 Miles | 2–4 Miles | 4–6 Miles | Over 6 Miles |
| Yes | 34 | Yes | 80% | 48% | 16% | 4% |
| Uncertain | 30 | Uncertain | 16 | 32 | 40 | 20 |
| No | 36 | No | 4 | 20 | 44 | 76 |

C.   PERCENTAGE OF "YES" RESPONSES: Within the 75% line, the "Yes" responses represent 75% or more of the total responses; 50%–74% "Yes" responses between the 75% and 50% lines; and 25%–49% between the 25% and 50% lines.

### Figure 4.5

Cartographic and Statistical Presentation of Responses.

and design of the questionnaire, the manner of its administration, the sampling procedures to be used, and the ways in which the data obtained are to be tabulated and analyzed. Each field research problem is unique, and the overall interviewing design must be developed exercising the greatest care to assure that pertinent data are collected as accurately and efficiently as possible.

## SUGGESTED REFERENCES

BABBIE, E. R., *The Practice of Social Research*. Belmont, California: Wadsworth Publishing Co., 1975.

DAVIS, J. A., *Elementary Survey Analysis*. Englewood Cliffs, New Jersey: Prentice-Hall, 1971.

GORDEN, R. L., *Interviewing: Strategy, Techniques, and Tactics*. Homewood, Illinois: Dorsey Press, 1969.

HIGHSMITH, R. M., JR., "Suggestions for Improving Geographical Interview Techniques," *Professional Geographer*, January, 53–55. Washington, D. C.: Association of American Geographers, 1962.

KAHN, R. L. and C. F. CANNELL, *The Dynamics of Interviewing*. New York: John Wiley & Sons, 1967.

KNIFFEN, F., "The Tape Recorder in Field Research," *Professional Geographer*, January, 83. Washington, D.C.: Association of American Geographers, 1962.

OPPENHEIM, A. W., *Questionnaire Design and Attitude Measurement*. New York: Basic Books, 1966.

PAYNE, S. L., *The Art of Asking Questions*. Princeton, New Jersey: Princeton University Press, 1951.

RICHARDSON, S. A., B. S. DOHRENWEND, and D. KLEIN, *Interviewing: Its Form and Functions*. New York: Basic Books, 1965.

STODDARD, R. H., *Field Techniques and Research Methods in Geography*. Dubuque, Iowa: Kendall/Hunt Publishing Co., 1982.

WARWICK, D. P. and C. A. LININGER, *The Sample Survey: Theory and Practice*. New York: McGraw-Hill, 1975.

WARWICK, D. P. and S. OSHERSON, *Comparative Research Methods*. Englewood Cliffs, New Jersey: Prentice-Hall, 1973.

# Sampling Procedures Appropriate for Geographic Field Study

## Chapter

# 5

**Chapter 5**

Geographic field investigations can encompass large areas as well as areas that have a wide variety of spatial phenomena. These may range from multistate regions to small, dense urban complexes. In both situations it is not possible to comprehensively map or investigate every occurrence of a phenomenon within the region. It is too costly and time-consuming to interview every urban resident. Therefore, to bring problems down to more manageable proportions, a small percentage, or a *sample*, of the total area must be selected for study. This sample is drawn from the larger body of data, which is called the *population*. The term *population* does not in this sense refer to numbers of persons, but rather to the total data within the research area.

For example, in attempting to determine the attitudes of urban residents toward a new shopping mall, the researcher must select a representative sample of persons from the total urban population. This sample could be representative in a number of ways. It could be drawn in a spatial manner in order to adequately represent all geographic regions of the city, or all socioeconomic classes might be represented equally, or all education levels, or political affiliations, or combinations of these. The end result would be the documentation of representative viewpoints toward the proposed shopping mall based on some selective criteria. In geographic field sampling, the emphasis is placed on spatial sampling.

## TYPES OF SPATIAL SAMPLING

Four different types of spatial sampling may be identified based on the intensity of the sample. These are:

1. Exploratory sampling,
2. Reconnaissance sampling,
3. Extensive sampling, and
4. Intensive sampling.

### Exploratory Sampling

*Exploratory spatial sampling* would be used in regions where little information concerning a particular phenomenon or phenomena exists. In such situations, the proper classification schemes, mapping keys, and other basic data might not be available, or have not as yet been documented. Exploratory spatial sampling would be designed to provide such information, leading to a more intensive sampling to be done at a later date. Most frequently, ground control or location is a problem in exploratory work, and providing spatial control in itself constitutes a valid goal of exploration.

Exploratory spatial sampling might also be used to determine whether or not seemingly homogeneous units of some phenomenon or groups of phenomena had, in fact, minor variations which previously were overlooked. Often, a lack of such information prevents the development of an adequate research design. When a field researcher is faced with such a situation, there are two methods of approach: (1) Only those features that might be relevant to detect possible subvariation of a phenomenon would be measured; or (2) a comprehensive documentation of every conceivable factor would be made. The latter would be time-consuming and might only be of value in later analyses and sampling, after the exploration stage, when systematic coverage might prove to be important. Some examples of exploratory spatial sampling would be sampling of unknown units of vegetation or the determination of broad-scale regional soil maps for areas in which none previously existed, or documenting crops and agricultural yields where no data were previously available. Thus, the chief purpose of exploratory sampling is to permit the actual mapping of phenomena where none were documented before and to organize information essential to a given problem. It is better to have a crude map of something than no map at all. Once different types of phenomena, characteristic areas, or classification units can be demonstrated, the critical phenomenon or phenomena may be isolated for further study and sampling.

### Reconnaissance Sampling

*Reconnaissance spatial sampling* is similar to exploratory sampling except that it is carried out in an area or region where more data and documented materials exist. Aerial photography, together with topographic or other map coverage, provide some ground control. Thus, in reconnaissance sampling, the organization and execution of the design may be more systematic than in exploratory sampling. Reconnaissance sampling may be conducted by air, car, or on foot depending on the size of the study region and the base map scale. For reconnaissance information to have maximum utility, the researcher must be aware constantly of the scale of the region and the minimum area considerations as discussed in Chapter 2. Generally, the base map scale should be two to four times larger than the final informational presentation scale. Reconnaissance spatial sampling is designed to cover a maximum area in a minimum time at a minimum cost with a maximum data return. This type of sampling may be the only sampling form suited to large regional-level projects, such as the location of a new settlement or the selection of possible routes for a transmission line network.

## Extensive and Intensive Sampling

There is no sharp boundary between *extensive spatial sampling* and reconnaissance spatial sampling. The principal difference is that with extensive sampling, there is enough existing background data to formulate a comprehensive research design. With extensive sampling, usually there is less generalization and abstraction of data, and minimum areal units are smaller.

Extensive spatial sampling is differentiated from intensive spatial sampling in that the former tends to relate to regional phenomena whereas the latter is more localized and perhaps represents a unique view of conditions at a particular site. In the measurement of attitudes, for example, the extensive spatial sampling might deal with generalized perceptions and opinions concerning a broad land-use problem, while intensive sampling might investigate the range of individual opinions in a specific neighborhood regarding a local and particular land-use issue. In the case of vegetation, as an example, extensive sampling would reveal different types of vegetation differentiations, whereas intensive sampling would reveal individual site variations and how this information relates to the resulting classification system.

In terms of scale alone, there is no clear demarcation between extensive and intensive spatial sampling. A researcher who deals with regions at a scale of 1:250,000 might feel that a scale of 1:24,000 is quite intensive, whereas a person who normally works at scales of 1:24,000 might consider 1:24,000 an extensive scale and 1:5000 intensive. In all cases, scales of about 1:5000 or larger are considered intensive.

Extensive and intensive spatial sampling also could be differentiated by the percentage or the degree of representation of a given sample. A very large percentage sample used to obtain uniform areal coverage over a widely diverse region would be considered intensive as opposed to extensive sampling where the numbers of observations would be considerably smaller, due perhaps to a lessened requirement for accuracy, or to employing a different set of phenomena criteria.

## SAMPLING UNITS

In addition to the general types of spatial sampling, it is possible to distinguish several different units of samples:

1. The point sample;
2. The area sample, or quadrat;
3. The linear sample; and
4. Plotless units, where no area is directly associated with a particular point but data are gathered throughout a zone close to the point location.

## Point Sampling

The point sample unit is a one-dimensional observation of what exists at a specific geographic location. A point sample is a point at all scales and does not vary in size with the scale of presentation or the scale of the overall sampling procedure. Mapping and surficial display based on specific point measurements are more acceptable conceptually than the general practice of summarizing data at points for mapping purposes. A great deal of secondary data, such as census information, is treated incorrectly in this latter fashion.

Several examples of point sampling serve to illustrate its use. In a land-use context, the type of land use observed at the particular location would be recorded. In a vegetation sampling context, a pointed rod might be dropped into a grassland area, and all those plants touched by the rod recorded. In the latter case, the rod might be supported from an elaborate *sample frame*. During a point sampling, an observation might fall on a transition zone, the boundary between two fields, for example, or the property line between an industrial site and a residential area. The resolution of such a problem would depend on the philosophical basis of the research design. In some philosophies, a resampling to obtain a "pure type" (homogeneity philosophies) would be in order. Other philosophies would consider rejection of the sample point a serious biasing of the data. Such problems should be resolved on a project-by-project basis.

With point samples, as with all samples, the question of location arises. A researcher may be able to locate adequately a particular sample point on an aerial photograph or on a topographic map and yet not be able to adequately relate this to a precise location in the field. The techniques of resection, triangulation, and intersection discussed in Chapter 3 may be useful in these cases. Depending on the degree of accuracy that is required and the scale of the aerial photograph and/or base map that is utilized, the researcher may be forced to accept a certain amount of locational error. Most often, the aerial photograph is the most desirable base map in these situations because, as indicated previously, an almost infinite number of visible orientation points exist. However, even aerial photography has its limitations, particularly at smaller scales where ground detail is poor and locational errors may result. The uniform training of field personnel in ground location of points and the definition of the minimal degree of locational accuracy acceptable are necessary for the proper execution of the sampling design.

The difficulty of applying keys and classifications developed in the office to field situations is a common problem in all types of field work. The uniform interpretation of keys and consistent estimation and documentation of field phenomena must exist from one field researcher to another. Great care must be taken to insure that the classification schemes do, in fact, work in field situations. An exploratory, or perhaps a reconnaissance, sampling may prove to be invaluable as a test to determine whether or not it is possible to make measurements in the desired way in the field. Finally, in all types of sampling, close liaison with field personnel will confirm or deny the adequacy of the sample in the light of general observations made on-site.

### Area Sampling

There are two variations of the area or quadrat sample: the fixed quadrat and the variable quadrat. If the sampling area remains constant and the shape of the quadrat uniform, it is a fixed quadrat type. If the size and/or shape of the sampling unit varies, it is a variable quadrat.

**Fixed Plots.** When dealing with *fixed quadrats*, the researcher must consider the factors of shape, size, orientation, degree of symmetry, and plot density. Circular and rectangular plots are the most commonly used shapes because of the ease of duplication in the field. For small sampling units, circles can be defined by using a string or a tape to swing a radius of the appropriate dimension. Similarly, a heavy cord with knots at the corners can be stretched between stakes or held by individuals to give a square or other rectangular shape. Both of these methods run into difficulty in heavy timber and rough terrain. In large sampling units, circular or rectangular plots may be defined on the base map.

Symmetrical plots should be used unless there is reason to purposely change the symmetry for a particular field study. For example, with a relatively uniform or homogeneous vegetative cover, land use, or soil condition, a symmetrical plot or quadrat is the most efficient. However, the researcher sampling phenomena that display a parallel arrangement—drainage patterns, for example—might find it advantageous to align nonsymmetrical quadrats so that the long dimension runs across the stream (a long rectangular shape might suffice). This would allow the zones on either side as well as the stream itself to be represented. The researcher then must be willing and prepared to accept the diversity that will exist between the stream bottom and the slopes of the stream valley. The same approach could be employed for sampling along a highway if in-depth information along both sides of the road was desired.

In considering the shape and size of quadrats to be used in the field and their orientation, it is sometimes valuable to field researchers to simulate the shape, size, and orientation of quadrats in the laboratory. This allows direct observation of sample effects. Sampling boards, both random and nonrandom, are available commercially. These are equipped with color spots pasted on plastic to give a repeatable population from which to sample. Such boards have documented statistical parameters, percentages, densities, etc., to provide the user with information as to what sort of data variation will result from manipulating the plot (see Figure 5-1).

The size of the plot is to a large extent determined by the density, or *grain*, of the phenomenon in question. A very dense or fine-grained phenomenon could be represented adequately with small quadrats. However, phenomena that are coarse grained or very heterogeneous would require a much larger quadrat to obtain the same information. In both situations, the researcher must determine what sample size is most compatible to a given situation. Two similar methods may be used:

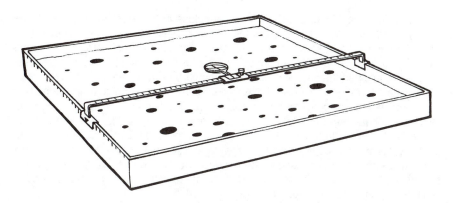

**Figure 5-1.**

A Sampling Board with a Random Population.

1. The optimum quadrat size may be determined in the field prior to sampling; or
2. A convenient small sample area may be selected, and a determination made of the number of times this quadrat must be "placed" in similar environments.

The latter approach may be employed prior to or during the sampling. Quite often, the best approach is to experiment with the size of the quadrat prior to sampling. In this way, an optimum size is immediately employed and less experimentation in the field is needed. A technique widely used in vegetation sampling is to employ the *species-area curve* which, when applied to other subjects, may be called the *phenomena-area curve*.

The phenomena-area curve is simply a graph of area or numbers of quadrats versus the number of species, house types, or other phenomena that are to be studied in the region. The area is usually plotted on the *x*-axis, and the phenomena on the *y*-axis. For a more or less uniform region, experimentation may determine the optimum quadrat size for accurately characterizing that region. For example, in a vegetation mapping problem, if a relatively fine-grained grassland is to be sampled, a meter-square quadrat could be used as an initial point from which to expand. The species within the quadrat would be identified, and then the plot would be enlarged by doubling its size. The enlargement would occur as shown in Figure 5-2A. The species found within the enlargement then would be listed. The area of these two quadrats combined is now 2 square meters. It is then further enlarged by doubling to create an area of 4 square meters. Once again the species would be enumerated. This area of 4 square meters would then be doubled to 8 square meters, and an enumeration of species made for the added area. After each enumeration, the *total* number of species in the entire sample area up to that point and the total area would be entered on the graph.

The field researcher would quickly notice that the number of new species acquired within this homogeneous grassland environment initially was high, but began to decrease rapidly as more areas were included. Eventually, a point would be reached where additional increase in area would not result in the addition of more species (see Figure 5-2B). A point very close to this leveling-off position and slightly downslope would represent the optimum sample area. In this example, where the vegetation has a similar grain size, a quadrat of 6 square meters is the most desirable. Alternatively, a convenient size quadrat is selected, and a species-area curve or phenomena-curve procedure could be employed to determine when an adequate number of samples have been taken within a particular location. This would be more time-consuming, however, than the former method. Beyond the method of phenomena-area curves, there seem to be no appropriate formulas to determine the correct number of spatial samples. In certain disciplines dealing with nongeographic data, appropriate formulas have been developed. However, at this time, such formulas do not exist for geographic data. In view of this lack, authorities suggest that samples of 10 percent to 20 percent are usable, and that samples of 30 percent are usually adequate, for geographic work.*

The field geographer should be aware that sophisticated variation in the conceptualization of sample areas and quadrats can exist. Multiple levels or scales of area samples may be simultaneously incorporated. For example, an area may be selected for sampling that has within it additional area, or other types of, samples. The choice of such nested sampling is, of course, directly related to the research and sampling design selected for the specific field problem. A similar graph of phenomena-location curves may be constructed for point samples when multifeature mapping is used.

**Parcelle Mapping.** A variation of the use of area or quadrat samples is in situations where the quadrats themselves actually touch edge to edge over the entire study area. The regular matrix of cells that results is called *Parcelle mapping.* The terms *isonome studies* and *cellurization* also refer to the same technique. The cell sizes are set to represent minimum resolution units as discussed previously. In Parcelle mapping, to define a manageable "sample" of the data population, a *generalization* is made based on the minimum scale of resolution (cell sizes) rather than on the basis of a percentage being sampled. The phenomenon is "abstracted" into these uniform mapping units. Since the procedure involves the entire region, it is technically not a sample, but rather a generalization. However, if aerial or other reconnaissance is employed to generate such data, the field check on-site may then constitute a sample. In the strictest sense, data generalization in Parcelle mapping is not to be considered as a sample, but it is relevant to the coalescence of fixed areal units. For example, if crop type mapping is desired, each cell is coded with the extant crop. If one type of crop fills a cell, there is no problem; however, if more than one crop type is present in any one cell, difficulties arise. Under a generalization or Parcelle mapping scheme, two solutions to the problem may be employed:

1. The crop having the greatest area within the cell may be listed; however, this procedure may not be acceptable from a philosophical point of view; and
2. It may be possible to determine all permutations and combinations of agricultural activities prior to the data gathering. For example, if an area had four crops—cotton, sorghum, corn, and alfalfa—there would be ten possible per-

---

*L. J. King, *Statistical Analysis in Geography.* Englewood Cliffs, New Jersey: Prentice-Hall, 1969, pp. 75–76.

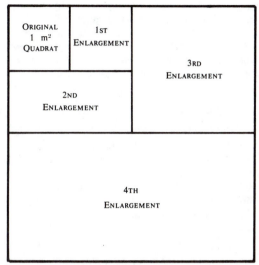

A. Experimental Sampling for Producing a Phenomena-Area
(Species-Area) Curve.

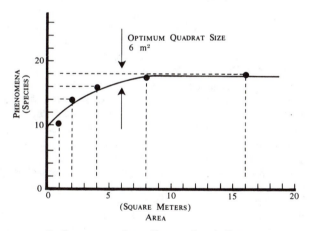

B. Phenomena-Area (Species-Area) Curve.

## Figure 5-2.

Experimental Sampling for Producing a Phenomena-Area (Species-Area) Curve.

mutations or paired combinations: (1) all cotton, (2) all sorghum, (3) all corn, (4) all alfalfa, (5) cotton and sorghum, (6) cotton and corn, (7) cotton and alfalfa, (8) sorghum and corn, (9) sorghum and alfalfa, and (10) corn and alfalfa. Each permutation could then be recorded for each cell. What actual permutations might exist in a given area could be determined by a previous exploratory sample, and the appropriate number of permutations would be used (see Figure 5-3).

Another type of Parcelle mapping consists of the estimation of the percent of a single phenomenon. Single factor maps would result from such mapping. For example, the percentage of land in escarpments, or the percentage of land associated with drainage patterns that are obviously damp and possibly subject to flooding, might be of value in locating a particular townsite. This form of data reduction complements secondary sources in that existing maps may be recompiled into cellular form and thus become compatible with the survey units employed for the field phase of the research.

Many types of land use and other questions may be answered in a practical sense by field research that does not require the intensity and detail of information that might be necessary in an academic work. A Parcelle mapping may be used, for example, to determine the location of a new townsite. It will provide an adequate data base when the study area is large. For example, using cells of one square mile in a matrix of 3000 or 4000 square miles would be adequate for the decision being made, since it would constitute an adequate resolution unit to select the general area in which the new town should be located. In many forms of field work, data are often taken at several scales of resolution, and because of benefit/cost and time considerations, it frequently is necessary to use generalization-level detail for entire regions. Potential sites selected at this scale can be intensively sampled and studied at a subsequent phase of the research.

Parcelle mapping may also be effectively used for regional data gathering using an airborne platform. When this is done it is called cognitive composite mapping.* This is an efficient data gathering technique that employs the cognitive abilities of the research geographer. The research geographer must first look at the study area from the aircraft and compare topographic map information with observed landscapes. As the aircraft flies definite patterns parallel to the parcelle mapping polygons (typically sections), the observer orients himself to these lines as they appear on the landscape. These minimum areal mapping units may be poorly defined on the ground and in some cases the polygons chosen will not be expressed on the ground; nevertheless, it is possible to reference these units effectively.

By concentrating on these polygons the observer can build a cognitive composite picture of both ground data and base map information. This is similar to the cognitive "picture" that an air traffic controller generates. Parcelle marking, however, generates a visual image rather than a mental one.

Once the researcher acquires the "picture", the composite image is so detailed it is possible to determine where even minor roads deviate from the

---

* A detailed discussion of the cognitive composite mapping procedure may be found in: F.T. Aldrich's "Real Time Airborne Data Acquisition and Polygonial Generalization for use in Spatial Synthesis Scenarios," Vol. 6, 18–27. *Papers and Proceedings of Applied Geography Conferences*, Toronto, 1983.

The grid figure showing the parcelle mapping with coded cells:

| | 3 | | | | 2 | 2 | 2 | 4 | 1 | 1 | |
|---|---|---|---|---|---|---|---|---|---|---|---|
| 3 | 3 | 3 | 3 | 4 | 2 | 2 | 2 | 4 | 1 | 1 | |
| 3 | 3 | 3 | 3 | 4 | 2 | 6 | 2 | 4 | 1 | 1 | |
| 1 | 1 | 5 | 5 | 6 | 6 | 6 | 6 | 4 | 1 | 1 | 1 |
| 1 | 1 | 1 | 1 | 5 | 3 | 3 | 5 | 1 | 1 | 1 | |
| | 1 | 1 | 1 | 1 | 5 | 3 | 3 | 5 | 1 | 1 | 1 |
| | 1 | | 1 | 5 | 3 | 3 | 3 | 5 | 1 | 1 | |

RESEARCH
AREA
BOUNDARIES

☐ MINIMUM AREA

 Sample quadrats used in this example to determine the classification categories prior to the generalization of Parcelle mapping

CLASSIFICATION/CODING

1. Agricultural
2. Industrial
3. Residential
4. Agricultural-Industrial
5. Agricultural-Residential
6. Industrial-Residential

**Figure 5-3.**

Example of Parcelle Mapping.

The size of the quadrat represents the minimum size area for mapping purposes. Each quadrat or cell is coded as to its dominant use. In this particular example, 75 percent or more defines dominant use. The grid represents minimum area cells, which could be existing sections, for example, or any grid devised by the researcher. The grid does not need special alignment or orientation—no bias will be introduced.

topographic map alignment, where a new building exists, or a former one was removed. Subtle differences in topography, drainage, and changes in vegetated areas (if originally mapped) are easily seen. The image follows the motion of the aircraft platform and remains registered at all times.

In practice it is convenient for the researcher to sit in the forward right-hand seat of the aircraft and tape the map underneath the side window with the polygons adjacent to the window and oriented with the line of view. The eye may then move rapidly from the map to the scene and back as needed. When the cognitive picture appears, two polygons (sections) may be handled at once, one on top of the other (background and foreground), as the platform moves past. Since the plane flies at eighty to one hundred miles per hour and at a height of five hundred to eight hundred feet, the researcher has about forty-five seconds to tape record information for both cells. This is more than adequate time as practice considerably speeds up the process. The tape recordings are later transcribed and compiled into topical maps and computer coded input. Other researchers have access to them throughout the project should they need additional perspective and landscape annotations.

Of course the researcher must be comfortable in an airborne environment to employ this method. The uninitiated may experience significant disorientation, and must concentrate to produce the cognitive composite view, or to maintain it for any length of time. However, with practice, it becomes a very efficient and cost effective field technique.

Types of data generated in this manner may include land use, land cover, topography, natural hazards identification, settlement patterns, drainage way location, and classification and mapping of visible industrial, mining and agricultural activity. The cognitive composite procedure is particularly cost and time effective when used in abstracting regions exceeding several thousand square miles in area. The procedural alternatives, air photography or remote sensing, are frequently more costly due to time delays in image acquisition and lack the asset of having researchers visually examine the entire spatial matrix from close range. This aerial perspective alone is often significant in the overall viewpoint and execution of the project.

**Variable Plots.** The alternative to fixed-plot sampling is the *variable plot*. In this form of sampling, both the size and shape of the plot may change from one observation to another. For example, Sample Plot No. 1 might be an irregular shaped area of 2.8 hectares, Sample Plot No. 2 a rectangular area of 0.6 hectares, and Sample Plot No. 3 an almost circular area of 5.0 hectares. These plots are called *relevés* or *ocular plots*. This form of sampling is used with research designs based on the philosophies of homogeneity and discreetness.

If we assume that groupings or classes of spatial phenomena exist, and that they have discrete boundaries and are made up of similar subunits which are distributed over the landscape, and if such complexes of phenomena are repeatable and easily recognized from one place to another, then it is possible to utilize these homogeneous areas to establish quadrats. For example, in vegetation studies, an area of homogeneity may be identified on aerial photography. This procedure is called *prestratification* and consists of selecting those areas having uniform tone, texture, pattern, shadow pattern, color, and combinations thereof, indicating some degree of homogeneity. In the field, the researcher then walks through the area and identifies those parts that are located well away from transition zones and records data from the entire homogeneous area. Each patch of repeatable homogeneity is considered to be a separate quadrat or relevé. In this form of sampling, large areas may be handled easily, and the time cost per unit of sample is low. Usually, data generated in this way must be analyzed by using presence-absence, tabular analysis, or some form of statistical procedure that provides a variable plot basis. The basic assumption is that if a given phenomenon is uniform, there is no reason to take samples that fall into areas that represent transitions between homogeneous units. Thus, it is not necessary to have large, dense sample networks to statistically eliminate nonuniform and transition areas.

Often this form of sampling is modified to conform with statistical methods that require randomness and uniformity in plot size. In this modified form, the revelé becomes a higher order sample area which is then sampled by using conventional point, area, or other sampling methods.

### Linear Sampling

*Linear sampling* procedures consist of one- or two-dimensional transects extending point to point (see Figure 5-4A). This is a widely used form of sampling because it is fast, simple, direct, and may be used along roads.

When the transect is one-dimensional, any phenomenon cut by the line will be recorded, and the numbers of such encounters can be used for frequency or density calculations. Also, percentages of the length of the lines that lie within various land uses, cover types, or crop categories provide a mechanism to characterize regions.

A convenient way to construct such transects is to use roads. In some cases, it is not possible to use the road as a true one-dimensional transect because the road may be the dividing line between land-use types or other phenomena. This may not be true, for example, with soils, but since it is still not possible to see the soils under the road, a similar problem exists. Also, the mere presence of a road can affect the data being gathered and bias the information being collected. These difficulties may be partly resolved by using a transect line that parallels the road instead of the road itself.

A second type of linear sample is the two-dimensional transect. In this procedure, an area adjacent to the transect line as well as larger zones on either side of the line may be studied. Again there are difficulties if the roads themselves are used. For example, in an agricultural environment, it may be possible to see as much as a quarter section on either side of a rural lane; however, in a built-up urban environment, often only the store fronts facing the street may be visible, and the researcher will have no knowledge of what occurs along the alley or in the middle of the block. Such problems may be resolved by using supplemental aerial photography. Even this may not completely eliminate the problem when transects are employed, because ground-truthing the image components may be difficult due to access problems, particularly if the transect lies off of a road. The area transect is valuable, however, because a specific percentage sample of the total study area may be selected for study (see Figure 5-4B). Such percentage areas may provide valuable insights for transition zones within the study region but normally will not provide adequate data for an in-depth understanding of any one transitional situation.

A variation of the two-dimensional transect consists of the placement of regularly spaced conventional quadrats along the transect lines. This allows for many standard statistical procedures to be employed in the analysis of the collected data. For example, assume that the problem is if and how perception of noise and air pollution varies spatially with distance from an airport. The relevant data must be collected by interviewing the residents of the research area. Perhaps there are several thousand households within the research area and time restraints restrict full coverage. A sampling procedure must then be constructed to reveal areal differentiations. One method might be to place points one mile apart along a transect line. In the field, the ten nearest households to a given point can be determined and, using

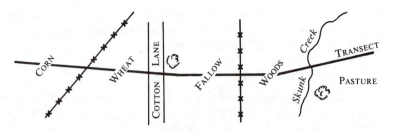

A. One-Dimensional Sample.

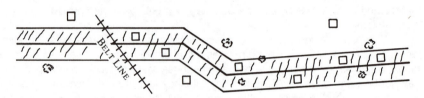

B. Two-Dimensional Sample.
   Transects should be selected to provide a representative cross section of the total research area. Generally, in an area of 250–300 square miles, a well-chosen transect of about one mile in width or ten to twelve sample areas of about one square mile in size should provide sufficient data to formulate generalizations and an overall perspective of the total area.

C. Conventional Quadrats Located along a One-Dimensional Transect.

D. Point Sampling along a Transect.

## Figure 5-4.

Linear Sampling Types.

118

a table of random numbers, one household in each group of ten can be chosen to be interviewed. This mechanism might reveal that perceptions change or that the intensity of perceptions vary with distance (see Figure 5-4C).

Point samples along a line also can be frequently used. One example of this is vegetation sampling using a rack of pins which are placed at points along a line. The pins are dropped to the ground, and the plants that the pins touch are then recorded. Similar point sampling in land-use determination or other sampling projects may be employed (see Figure 5-4D).

## Plotless Sampling Units

The *plotless unit* is a point sample that is associated with an undefined area. Plotless sample methods had their origin in vegetation sampling in Wisconsin during the early 1950s. They may be applied to many sorts of phenomena. The basic method has four variations each employing only distance measurements. These distances are obtained either from the sample point to a phenomenon or between phenomena. The points must be randomly allocated, or at least a random component must be a part of the sampling scheme (typically, points are located along a line transect). If the mean (D) of the individual distances (d) collected over the total sample is multiplied by a constant (K) the number becomes a density value. This constant varies from 0.8 to 2.0.

Thus, with ten thousand representing the number of square meters in a hectare, the general formula may be expressed:

$$\text{density per hectare} = \frac{10,000}{K (D \text{ in meters})^2}$$

In this instance, all distances must be measured in meters. If the researcher desires to work in acres, then the numerator becomes 43,560 (the number of square feet in an acre), and the distance measurements must be made in feet.

In the first variation, the *closest individual method,* the distance from the sample point to the closest individual or phenomenon of interest is recorded irrespective of the direction (see Figure 5-5A).The value of K in the formula is 2.0

The *nearest neighbor method* records the distance between the closest individual and the "neighbor" of the same type closest to it. (See Figure 5-5b). Theoretically, the value for K should be 2.0 also; however, because of difficulties in obtaining true random pairs, a working value of K = 1.67 should be used.

The third variation is the *random pairs method.* In this variation the nearest individual or phenomenon type to the sampling point is used, and centered around this individual with the vertex at the sample point, a 180 degree angle of exclusion may be defined. The distance from this individual or phenomenon type to its nearest neighbor outside of the exclusion angle is measured (see Figure 5-5C). This variant requires a value for K = 0.8.

The final variation, the *point-center quarter method*, calls for the researcher to draw lines through the sample point in the cardinal compass directions—north, south, east, and west—thereby dividing it into quarters. Within each quarter, the distance from the sample point to the nearest individual or phenomenon is

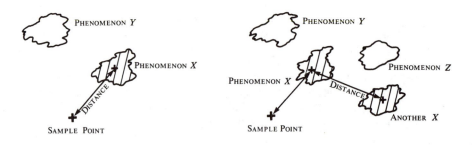

A. Nearest Individual.     B. Nearest Neighbor.

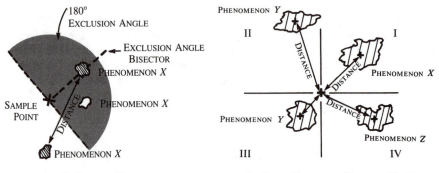

C. Random Pairs.     D. Point-Centered Quarter Method.

**Figure 5-5.**

Plotless Sampling.

measured, providing four measures for each sample (see Figure 5-5D). Here, the K value for the formula becomes 1.0. This variation supposedly has the greatest reliability.

Each of these variations is necessary to obtain information for all categories of land use, species, or classes of phenomena at each point. A table may then be constructed with the "columns" representing randomly selected data points and the "rows" representing types or classes of phenomenon. The distance measures are entered in each cell of the matrix respectively.

For the point centered quarter method four distances will appear in each cell, to be averaged before creating means derived from the total sampling. These overall means are calculated by summing each row and dividing by the number of points sampled.

Combinations and variations of these basic plotless methods can be employed easily in field research. The basic assumption of this type of sampling is that there are no discrete units or spatial phenomena and that there are only gradual changes from one area to another; i.e., a gradient is involved. This type of sampling is opposite in philosophy to the homogeneous approach. This technique can be applied to any geographic field measurements in which the phenomena cannot be classified into distinct units. An example of such a continuum might be a very broad transition zone such as the gradation from urban into rural land use where instead of a sharp boundary there is a very wide zone or fringe several miles wide, possibly covering an entire portion of an urban area. The zone would have the characteristics of both urban and rural land use, which gradually would intergrade over distance. In the original application of the point-centered quarter method, random points were employed for locating the plotless samples. This may, however, be varied to fit geographically oriented reference frames.

## SPATIAL SAMPLING DESIGN

In addition to considering the type of spatial sampling and the units involved, the researcher must understand the theoretical basis of sampling design as it applies to locating the actual sites for geographic field measurements. There are three major procedures or approaches for placing the sample points, plots, or lines: (1) hierarchical sampling, (2) random sampling, and (3) systematic sampling.

### Hierarchical Sampling

In hierarchical sampling, data are taken from a variety of levels to gain an understanding of an entire process of hierarchical phenomena. For example, sampling might range from microsite local levels through regional samples, interregional samples, and perhaps samples on a state or nationwide basis. Samples might be taken within a county, within specified townships of that county, within individual farms, and within individual fields on those farms. In such cases, the actual selection of the exact points or areas involved are only guided by the hierarchy. The ultimate placement is made by employing one of the other approaches discussed below and the approach to use depends on the philosophy and purpose behind the sampling.

### Random Sampling

From a spatial viewpoint, a strictly random sample is not usually adequate. Randomness and completely unbiased samples are valuable in statistics, but in order to get adequate areal coverage within a particular region, some other method must be employed. The use of only random allocation will not provide even coverage. For example, if an area has an urban complex in one corner and the remainder is agricultural, the possibility of randomly selecting a representative sample of each land-use type and the transportation interconnection activities within the study area is unlikely (see Figure 5-7A). Unless the entire area is almost completely homogeneous, a different approach must be used. Therefore, random sampling must be employed in conjunction with either a systematic or some other form of stratified sample which provides adequate areal coverage. The exception to this rule would be the random orientation or transect direction for linear samples.

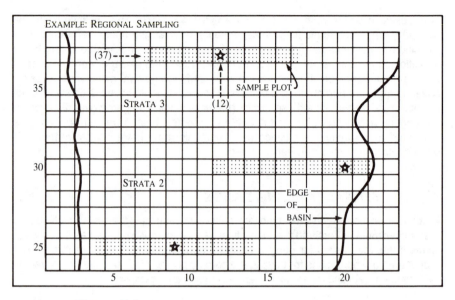

Figure 5-6.

## Systematic Sampling

If sample points are placed in a regular pattern, such as a matrix with rows and columns with the samples located at the intersections of each row and column, the sampling design is considered to be systematic (see Figure 5-7B). All points in this matrix are aligned from top to bottom and right to left. Such a design provides more adequate areal coverage than a strictly random one, but it also raises a problem. Any regularly occurring or periodic phenomenon might happen to have the same "period" as the spacing between either the columns of points or rows of points, assuming that the directional alignment of the periodic phenomenon is the same as the systematic sample (see Figure 5-7B). One method of countering this difficulty is to devise a systematic sample with a random start and perhaps even a random orientation. However, there is still the possibility that by accident a periodic phenomenon may be aligned regularly with the matrix.

An unaligned matrix is a better way to resolve the problem. In this type of sampling, often referred to as *stratified random sampling*, a network of grid squares is employed much in the same way as the grids for Parcelle mapping. The difference is that in stratified random sampling a sample point is located randomly within each grid square (see Figure 5-7C). Therefore, if there is a matrix of five rows and five columns of cells, each cell will have one randomly allocated point. This provides a systematically unaligned set of points and circumvents any chance of periodic phenomena falling exactly between the alignment of all the data points. This sample method is clearly superior to the systematic sample with a random start.

The stratified random sample, often called *Latin squares*, provides several ways for randomizing the observations within each cell. The use of a table of random numbers and an *x, y* coordinate axis for each cell is one device (see Figure

5-7D). A random selection of one initial pair of *x* and *y* coordinates for each row and column may be used in another method. In the latter case, the *x* distance of each point from the left margins of the cells within any one *column* is constant, and the *y* distance of each point from the bottom margin of each cell within a *row* is also constant. Then the combination of these distances for each cell in the matrix yields a random alignment (see Figure 5-7E).

In addition to spatial stratification, it is often useful to obtain samples that are not spatially stratified or that have stratification other than a spatial one. For example, if a researcher wanted to determine the attitudes of residents within an area toward a particular urban development, it would be necessary to adequately sample the area from a spatial standpoint, but it would also be necessary to get the viewpoints of various socioeconomic groups and education levels. Perhaps even further stratification based on different occupations would be necessary to get a true picture of the range of individual viewpoints within the research area.

Stratification for other phenomena or characteristics would entail the acquisition of additional numbers of observations within the appropriate population or other groupings. If spatial clusterings of such variation exist, it is possible, although not common, to combine the spatial with another stratification simultaneously. In complex sampling situations, the geographer who leaves the comfortable realm of spatial stratification should consult a statistician.

## Selecting a Sampling Design

Several types of sampling procedures and the basic principles of each have been discussed. In the final analysis, the scale and nature of the specific research problem will determine which of the many alternatives is most suitable. The researcher must keep in mind that sampling procedures are not the goal in themselves, but are the means to collect data essential to the problem, accurately and in the most efficient manner possible. Like other components of a research problem, such as the selection of base maps, mapping techniques, design and administration of questionnaires, the choice of a sampling mechanism and its degree of sophistication must be made after a careful consideration is given to all possible alternatives. The intellectual exercise of devising the research design to meet the needs of a given problem is a major part of geographic research.

## Two Practical Examples

**Hierarchial Sampling Example.** Hohokam River Basin Phreatophyte Study—plants that are heavy water users in the arid southwest. Basin characteristics include:

1. A north-south distance of 140 miles.
2. An east-west distance varying from thirty to eighty miles.
3. Stream segments generally running either north-south or northeast-southwest.
4. An elevation increasing to the north end of the basin.
5. Phreatophytes tending to be concentrated in the south end of the basin.

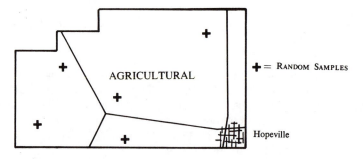

A. RANDOM SAMPLING. Random selection may not adequately represent the study area.

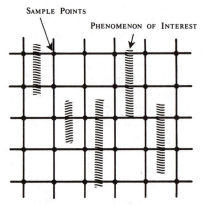

B. SYSTEMATIC SAMPLING MATRIX.

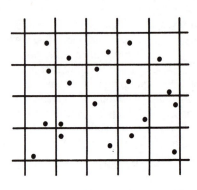

C. STRATIFIED RANDOM SAMPLING MATRIX.

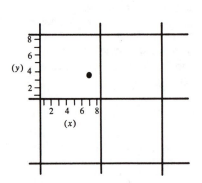

D. *xy* COORDINATES USED TO ALLOCATE POINTS.

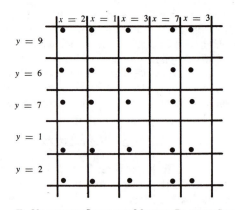

E. UNALIGNED SAMPLING MATRIX; RANDOM SELECTION OF ROW AND COLUMN VALUES ONLY.

## Figure 5-7.

Sample Placement.

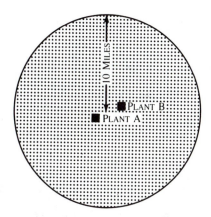

**Figure 5-8.**

Central Ohio Air Quality.

*Regional level sample:* The basin is divided into seven strata, twenty miles wide north to south. Within these strata fourteen regional strips (1 × 10 miles) are placed randomly using square mile cells. The strip is centered east-west over each chosen cell unless the selected cell is too close to the basin boundary to center the strip. In that case the strip is moved to the basin edge. Regional sample strips total 140 miles in length; each strip is examined on large scale air photography to identify possible phreatophyte stands where stream segments cross the strip (see Figure 5-6).

*Local sample:* Each prospective phreatophyte stand is sequentially numbered west to east on each strip, and using a table of random numbers, two sites are randomly selected on each. If all strips turn out to have suitable stands, the maximum possible field-visited stand measurements will be 28. At each selected site only the areas immediately adjacent to the stream channel will be investigated.

**Stratified Random Sampling Example.** Central Ohio Air Quality Perception Study: This region is the home of two large portland cement factories that are sources of particulate pollutants. The land use in the study region is mixed with large residential areas lying in close proximity to the plants. Respondents are residents of the area.

*The sampling scheme:* On an air photo mosaic base map of the area dots are placed at half mile intervals north-south and east-west for a radius of 10 miles around the sites. The ten houses nearest each dot are circled; these houses are sequentially numbered clockwise from the northwestern-most structure, and a random number table is then employed to select one of the ten houses for administering the sample questionnaire. The resulting sample size is 480 (see Figure 5-8).

## SUGGESTED REFERENCES

BERRY, B. J. L. and D. F. MARBLE, eds., *Spatial Analysis: A Reader in Statistical Geography.* Englewood Cliffs, New Jersey: Prentice-Hall, 1968.

COCHRAN, W. G., *Sampling Techniques.* New York: John Wiley & Sons, 1963.

GREER-WOOTTEN, B., *A Bibliography of Statistical Applications in Geography,* Commission on College Geography, Technical Paper No. 9. Washington, D.C.: Association of American Geographers, 1972.

GREGORY, S., *Statistical Methods and the Geographer.* London: Longmans Group, Ltd., 1968.

HARVEY, D. W., "Data Collection and Representation in Geography," *Explanation in Geography*, 266–281. New York: The Natural History Press, 1970.

KING, L. J., *Statistical Analysis in Geography.* Englewood Cliffs, New Jersey: Prentice-Hall, 1969.

KRUMBEIN, W. C. and F. A. GRAYBILL, *An Introduction to Statistical Models in Geology.* New York: McGraw-Hill Book Co., 1965.

KUCHLER, A. W., *Vegetation Mapping.* New York: Ronald Press Co., 1967.

MATALAS, N. C., "Geographic Sampling," *The Geographical Review*, Vol. 4, 606–608. New York: The American Geographical Society of New York, 1963.

SHIMWELL, D. W., *The Description and Classification of Vegetation.* Seattle, Washington: University of Washington Press, 1971.

STUART, A., *Basic Ideas of Scientific Sampling.* New York: Hafner Press, 1962.

THEAKSTONE, W. H. and C. HARRISON, *The Analysis of Geographical Data.* London: Heinemann Educational Books, 1970.

WOOD, W. F., "Use of Stratified Random Sample in a Land Use Study," *Annals of the Association of American Geographers*, Vol. 45, 350–367. Washington, D.C.: The Association of American Geographers, 1955.

# Field Research Design

## Chapter

# 6

# Chapter

All geographic field studies, regardless of the magnitude, must have a design or plan of operation. This design consists of four interrelated phases and serves as the basic framework within which the field research project is conceived and carried out. These basic phases are:

1. The preparation or planning phase in which all planning and prefield research is completed prior to the actual collection of field data;
2. The pilot study or reconnaissance, when the products of the planning phase are tested in the field;
3. The actual field data acquisition phase; and
4. The phase of analysis and presentation of the research findings (see Figure 6-1).

*Note that these phases which comprise the conceptual structure of field research design are basically the same regardless of the magnitude of the field research problem. The amount of time spent on each phase will vary from one research*

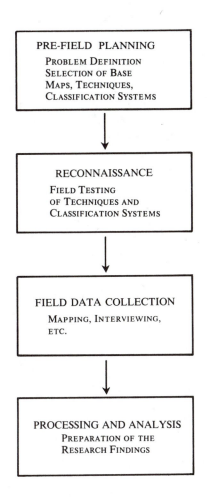

```
┌─────────────────────────────┐
│  PRE-FIELD  PLANNING        │
│  PROBLEM DEFINITION         │
│  SELECTION OF BASE          │
│  MAPS, TECHNIQUES,          │
│  CLASSIFICATION SYSTEMS     │
└─────────────────────────────┘
              │
              ▼
┌─────────────────────────────┐
│      RECONNAISSANCE         │
│      FIELD TESTING          │
│   OF TECHNIQUES AND         │
│   CLASSIFICATION SYSTEMS    │
└─────────────────────────────┘
              │
              ▼
┌─────────────────────────────┐
│  FIELD DATA COLLECTION      │
│  MAPPING, INTERVIEWING,     │
│  ETC.                       │
└─────────────────────────────┘
              │
              ▼
┌─────────────────────────────┐
│  PROCESSING AND ANALYSIS    │
│  PREPARATION OF THE         │
│  RESEARCH FINDINGS          │
└─────────────────────────────┘
```

**Figure 6-1.**

The Basic Phases of Field Research Design.

problem to another, and the complexity and details of each phase will vary enormously depending upon the nature and scope of a given problem.

In order to make proper and complete use of the powerful and sophisticated field techniques which exist today, the researcher must carefully organize the research effort from the problem statement through the planning, phasing, data acquisition, and analysis to the ultimate conclusions. Since the entire process is multiphased and dynamic, the field research design must be flexible and yet provide the needed rigorous scientific orientation. The multifaceted aspect and complexity of an adequate field research design are illustrated in the conceptualized flowchart shown in Figure 6-2. The complexity of the research design will vary depending upon the magnitude of the problem. In field research problems that are of long duration and involve numerous researchers, the research design may be highly elaborate and in practice will include all of the items shown in Figure 6-2.

For many projects, however, the problem magnitude will be small, and the workforce one or two individuals, with the entire project lasting only a few days.

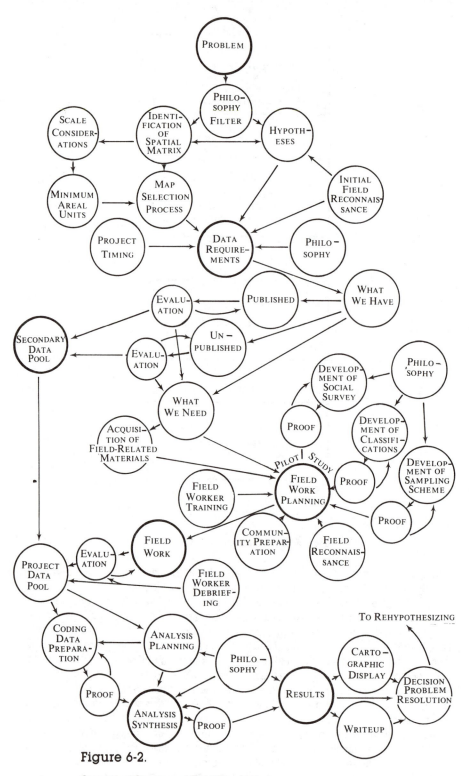

**Figure 6-2.**

Conceptual Structure of Field Research.

In such cases, the flowchart of the actual work would be highly abbreviated, omitting large portions of the proposed conceptual structure. For even such small projects, however, the researcher must consider, intellectually at least, all of the listed items during the design and planning phases of the research. Even though large portions of the illustrated flowchart may be dismissed after only a few moments of thought on a particular project on the basis of insufficient time, cost factors, or inappropriateness for some other reasons, it is absolutely necessary that the researcher have the opportunity to consider and reject such items.

Field research design and implementation is a highly creative process which is, in practice, a trade-off situation among the factors of cost, time, workforce, data volume, accuracy, and desired results. This compromise often represents the greatest challenge faced by the field researcher and may make the difference between successful project completion and frustrated failure. This is particularly true in contract, time-constrained projects; however, it is still somewhat the case even in purely academic questions approached with a more relaxed time framework and focus.

In a field research study involving only one or two researchers over a short time span, the planning phase often represents the majority of the total time allotted to the project. Although the problem is often relatively straightforward, decisions concerning the type and scale of data to be collected, selection of base maps, mapping and/or interviewing techniques to be employed, and the systems of classification to be utilized must be made with extreme care. Most important, the problem must be clearly identified. What specifically does the researcher wish to accomplish, and how does he or she intend to attain these goals?

The field testing of the choices made in the planning phase to confirm the workability of base maps, techniques, classification systems, etc., may be done quickly if the research area is small and the problem does not require a great deal of diverse data. Normally the number and range of alternatives to be field tested in a research problem of a small magnitude are few. If the planning and reconnaissance phases are completed with care, the actual field procedures should progress smoothly and without undue difficulty.

The final phase, processing, analyzing, and presenting the research findings, often is time-consuming. Frequently a project involving only one or two researchers is not equipped with the hardware facilities to expedite the cartographic and statistical analysis (see Figure 6-3).

## THE PROBLEM STATEMENT

The field research design must begin with some problem or question to be answered. The problem must be precise in its wording, and the terms fully defined. The problem might arise from previous research, or the researcher may identify gaps in subject matter, concepts, or methodology in the literature which justify additional research. The problem may be defined in general or specific terms if it is to be done for an employer, government agency, or private corporation. In some instances, it may be the response to a request for a proposal to solve a particular question. In any event, the problem will exist in one form or another. The researcher

| PHASE | TASK | PROPOSED TIME |
|---|---|---|
| PHASE I—PLANNING | Problem definition<br>Delimitation of research area<br>Assessment of existing information<br>Formulation of hypotheses<br>Selection of base maps<br>Development of classification systems<br>Development of data collecting<br>   techniques<br>Selection of sampling procedure | April 1–8<br>(8 days) |
| PHASE II—<br>RECONNAISSANCE/PILOT STUDY | Overview of research area<br>Field testing of classification systems<br>Field testing of data collecting<br>   techniques<br>Redefining of hypotheses<br>Preparing community/research area | April 9–11<br>(3 days) |
| PHASE III—<br>FIELD DATA COLLECTION | Mapping<br>Interviewing<br>Other data collection | April 12–22<br>(11 days) |
| PHASE IV—<br>PROCESSING, ANALYSIS,<br>PRESENTATION OF RESEARCH<br>FINDINGS | Cartographic analysis<br>Statistical analysis<br>Conclusions<br>Summary of research results | April 23–30<br>(8 days) |

**Figure 6-3.**

An Example of Work/Time Flowchart (A problem of one-month duration).

may have to restate the problem in understandable geographic terms or to limit its scope to one of manageable size. A field research problem implies acquisition of some new data obtained from observations made in the field.

## PHILOSOPHY OF THE RESEARCHER

The interpretation and understanding of any specific problem take place within the framework of the particular philosophy or viewpoint of the researcher. This philosophy acts as a filter to color the perceptions and nuances of the problem statement.

The philosophy also guides the interpretation, planning, analysis, and synthesis strategies employed throughout the research design.

A wide variety of philosophical viewpoints have been mentioned in previous chapters, some having wide scope and others very specific. From the standpoint of a generalized perspective, there are two viewpoints from which spatial phenomena may be perceived. First, the phenomena may be a part of some classifiable entity which exists and repeats itself over the landscape. Retail trade areas, for example, are a characteristic component of the classifiable phenomenon of urban land use. This classification type is repeated over the landscape and is homogeneous enough in its component characteristics to be recognized, classified, and mapped. Much of field work consists of identifying such units and mapping their boundaries. This represents the *discrete classification philosophy*.

Second, the phenomena may not be recognizable components of any particular unit or groupings. No distributional breaks may exist in any elements of the landscape, and changes only occur gradually through distance. Thus, there exist only very broad gradients or continua which do not lend themselves to identification as discrete units. This is the *continuum philosophy*. An example of such gradient phenomena might be the components of the rural-urban fringe mentioned previously which change gradually, often over large portions of an urban environment. These gradual changes are frequently identified as transition zones.

In reality, these two views differ only in scale. At a large enough scale, any classification unit may become a wide uniform gradient of conditions and phenomena. Alternatively, a regional scale continuum may be viewed at a very small scale, in which case it could be identified as a narrow transition between uniform classifiable units or could become a classifiable unit in itself. Under the appropriate scaled view, all continua become discrete phenomena, and all discrete phenomena can be viewed as continua. In geographic field research, since scale and uniform mapping units are of paramount importance, the researcher must differentiate phenomena resolutions. Thus, both viewpoints have a place in field methodology.

Other philosophies deal with random or nonrandom distributions over a landscape, the operation of stochastic processes to explain existing distributions, the assumptions of mini-max (minimum effort to maximize results) in a social or cultural setting, and so on. Regardless of the viewpoint, the philosophy acts as a filter through which the problem is interpreted. Also, the philosophy will influence the data requirements, sampling design, development of social survey instruments, classification development, analysis and synthesis processes, and final interpretation of results.

Two paths lead from the philosophical viewpoint (see Figure 6-2), one to the hypothesis formulation, the other to the identification of the spatial matrix or study area. Also, an interaction pattern exists between the two.

## THE SPATIAL MATRIX, SCALE CONSIDERATION, AND MAPPING

Each geographic field problem has associated with it a specific area, large or small, depending on the requirements of the problem. This research area may be continuous, as is the usual case, or discontinuous in the form of patches of a phenomenon or a disjunct distribution, such as might occur in mountaintop situations.

The spatial matrix must be defined clearly on a map for the field research design to be properly created and implemented. Without proper spatial delimitation, the field research would be unmanageable, and increases in cost and time due to extraneous samples would likely occur.

After the spatial matrix has been identified, a scale decision must be made. This is necessary so that the sizes of the minimum areal mapping units may be identified for the three types of mapping which may be employed during the study (see Figure 6-2). The three types are (1) the base map used during field data gathering, (2) the analysis or intermediate working scale maps used for data organization and display, and (3) the maps used for final data presentation. For each, the minimum area units, scale, and type of map used may vary or be uniform. As indicated previously, air photography is preferred for field base maps. However, for data organization and data presentation, planimetric mapping may be preferable.

The scale and size of the minimum areal units should be chosen to be cost effective and compatible with the philosophical viewpoint employed. The type of base maps used should be selected according to the discussion in Chapter 2.


## HYPOTHESIS FORMULATION

An hypothesis is an assumption, conjecture, or educated guess as to what the research might reveal. It is a hypothetical solution, or a theory yet to be proved, that offers a plausible answer to the research problem. The hypothesis and the problem statement by nature are closely related, and together they serve as the basis for, and establish the pattern of, the research investigation.

Not all research problems have distinct and separate hypotheses. The problem may be stated in such a way that it includes the hypothesis or hypotheses, so a separate statement would be of no value. In some situations, the researcher may not know enough about a given problem or research area to make a reasonable assumption as to what the study might uncover. However, the problems in geographic field research are such that sufficient information normally exist, or can be acquired, to formulate an hypothesis of some sort. It is obvious that the more information the researcher has about the area and problem, the more specific and credible the hypotheses.

The major purposes of an hypothesis are to further define the research problem and direct the research along the most potentially productive path. A sound hypothesis has several characteristics. It must bear directly and specifically on the overall research problem, and it should be capable of being phrased as a question. For example, if the hypothesis is that perceptions of noise pollution are associated with distance from the airport, it could be worded, "Do perceptions of noise pollution decrease with distance from the airport?" The hypothesis may be phrased in a negative way and be of equal or greater value than a positive question in some problems. In this example, the negative question, or *null hypothesis*, could be stated, "Perceptions of noise pollution are not associated with distance from the airport?" If the hypothesis is worded in the form of a question, it must be stated so that it can be confirmed, denied, or confirmed with reservations in light of the subsequent research. That is, it must be capable of being answered "yes," "no," or "perhaps." The question type hypothesis may be valuable in some problems as an aid in focusing the research efforts along definite paths and in eliminating, in part, the danger of tangential excursions.

In geographical studies, hypotheses are generally formulated in one of two ways. The first is concerned with exploring the causes of the spatial distribution of the phenomena studied in terms of the laws governing their locational behavior. In these cases, the researcher wishes to account for the where and why of the spatial distribution of a given phenomenon. This distribution is a *dependent variable* since its location depends on other factors. For example, in a problem concerned with land-use development in an area of rugged terrain, the location of the frost line at a given time may be an important aspect of the research. Knowing the effects of elevation and latitude, the researcher could hypothesize that the frost line would run along a given path. The researcher then would test this hypothesis in the field and find out how close the prediction was to actual conditions. Such types of hypotheses assume knowledge of established laws; in this example, the effects of elevation and latitude and, perhaps to a lesser degree, temperature inversions and air drainage.

The second method is by far more common in geographic field studies. It consists of the comparison of the spatial distribution of the phenomenon to be studied with the distributions of other supposedly related data. The spatial distribution of the dependent variable is first ascertained then compared with the distribution of other known and perhaps related phenomena. On the basis of these comparisons, an hypothesis could be formulated indicating that a spatial association exists. Research would reveal if such an association does indeed exist and whether or not it is a causal association. In essence, the researcher is analyzing a number of independent variables to determine if they can account for the distribution of the phenomenon being studied. For example, if the problem is, "Why does the spatial distribution of attitudes regarding a proposed new school bond vary so greatly?" the possible hypotheses might include family size, income, occupation, type of housing, as well as a large number of other variables. It is not likely that the researcher in this particular case will be able to fully account for the distribution of attitudes, but by selecting the more significant variables, he or she should be able to explain the pattern to a large degree.

As a component of the overall research design, the formulation of the hypothesis forces the researcher to think through and refine the research plan. It aids in sorting out the relevant from the irrelevant and narrowing the research efforts to concentrate on the critical issues. The hypothesis may be well defined or more generalized, and there may be one hypothesis or several, depending on the magnitude and scale of the problem and the researcher's knowledge of the area.

## THE RECONNAISSANCE

The primary objective of the reconnaissance is to obtain an overview or general perspective of the research area prior to the actual data collecting phase of the research. Ideally, the reconnaissance should cover the total research area or, at the minimum, ascertain the general characteristics of each of the subareas. In the later phases of the research, the reconnaissance should serve as a pilot study to test the research instruments previously developed. It is the time to try out the classification systems under actual field conditions, as well as the questionnaire, mapping systems, and other data collecting techniques. In addition, the hypothesis or hypotheses may be refined and the sample areas selected or, if the selection of sample areas has been made previously, confirmed or modified in the field. Also, the field work plan can be made to include a general time table and sequencing of the field work.

Whether or not a reconnaissance is necessary depends to a large degree on the size of the research area. If the area is small and the researcher is very familiar with it, the reconnaissance phase of the research plan may be waived. On the other hand, if the research matrix is large and the researcher does not have first-hand knowledge of all the complexities of the area, a reconnaissance is an essential part of the overall research design. If the researcher has doubts as to whether or not to invest time in completing a reconnaissance, it should be undertaken. It is difficult, time-consuming, and inefficient to change any aspect of the data collecting phase once it is underway.

## IDENTIFICATION OF DATA REQUIREMENTS

The circle below "Map Selection Process" in Figure 6-2 is titled "Data Requirements." This is an important nodal point in the research design. Here the results of the map selection process, hypothesis formulation, philosophical viewpoint, initial field reconnaissance, project timing, and management planning strategy are synthesized. This variety of inputs is necessary to define the data base requirements for answering the problem statement.

A comprehensive assessment and critical evaluation of all existing secondary sources of information pertinent to the problem is then made. The assessment and evaluation of secondary sources should at least include:

1. Scale of the information;
2. Dates when the data were obtained; and
3. The persons or agencies responsible for the data collection, compilation, and analysis.

These secondary sources include conventional published sources, such as census data, journal articles, soil conservation survey information, locally produced statistical and Chamber of Commerce types of documents, comprehensive plans, and others (see Appendix C.)

A wide range of unpublished data also are usually available for a particular study area. These may be located in state, local, regional, and federal government offices; private corporations; and within the files of local individuals and groups. Such information might be in the form of reports and maps and might constitute house documents of the agency or corporation; files of loose-leaf information (for example, forest fire occurrence reports held within the state foresty department, which contain dates, locations, and extent of burning); and so forth. Federal, state, and local agency field offices are usually fertile sources of unpublished, yet pertinent data. These agencies include: Bureau of Land Management, Forest Service, U.S. Department of Agriculture, Soil Conservation Service, Bureau of Indian Affairs, National Park Service, highway departments, fish and game departments, departments of revenue, school districts, fire protection organizations, water districts, and many others.

Much secondary unpublished information exists in manuscript form and only one copy is available. Some public relations expertise may be necessary to gain access to this information. This is particularly true in instances where intensive research efforts focus on certain regions. For example, in the developing coal mining areas of the western United States, many consulting firms and government agencies are doing research based on similar data requirements. Hundreds of researchers have descended on local offices in such regions looking for the same data. Manuscripts and maps are particularly vulnerable to destruction under such circumstances. In these instances, the researcher must be diplomatic and take the greatest care not to damage the documents, or the relationships with the personnel of the office. All forms of on-site recording media from cameras to tape recorders should be employed to ease the strain on the agency and the documents.

Some secondary information may be in the form of specifications and engineering data. For example, when a capital facilities study is part of the research effort, data on the diameter and flow rates for water and sewer systems, power substation supply potentials, and potential utility hookups of all types should be available at a variety of local offices. Data such as the number of students an elementary school can handle, or the service level that might be supplied from a five-officer local police force, usually are not secondary information and must be obtained through field interviews.

All secondary information must be pooled and carefully evaluated from the standpoint of the specific project to determine its adequacy. From this, a comprehensive listing of field data to be acquired on-site will be produced. The careful assessment of secondary sources of information will determine what additional data must be obtained from field observations.

## CLASSIFICATION OF DATA

The data to be used in the research problem, whether from secondary sources or from field observations, must be classified or categorized in some meaningful and workable manner. The classification systems devised are extremely important since this generalization process determines the detail or scale in which the data will be collected and sets the pattern for subsequent analysis. Many standardized classification systems exist for such phenomena as land use, soil types, vegetation, landforms, etc. These classification systems must be carefully assessed, and if they are adequate for a given problem, they should be employed. In this way, the findings of the specific research problem can be compared to the findings of other similar studies. However, if existing classification systems are not compatible to the given problem, new categorizations must be developed. The intellectual exercise of developing new or modifying existing classification systems is demanding and rigorous. It is a major component of the overall research problem.

## PROJECT TIMING AND MANAGEMENT

Project timing and management are important to all phases of the field research design system. A *work/time plan* should be designed to provide a framework within which the researcher(s) can organize time and resources. In all important tasks, it is essential to know exactly what the goal is, what mechanisms are to be employed in attaining the goal, and about how long it will take to complete the tasks. The work/time plan consists of a regular arrangement of steps, to be completed in order, a time schedule, and a list of materials necessary to complete the steps involved. The amount of detail of a work/time plan will vary greatly depending upon the magnitude and scale of the problem and the number of researchers and personnel involved in the project. A large-scale research operation lasting several months and involving significant numbers of researchers requires a highly elaborate work plan.

Regardless of the magnitude of the field research problem, a work/time plan should consist of four major and interrelated phases:

1. The planning phase,
2. The reconnaissance or pilot study,
3. The actual field data collecting phase, and
4. The processing and analysis of the data collected and the presentation of the research findings.

The planning phase is critical because it sets the pattern for the entire research project. It includes the definition of the problem; delimitation of the research area; evaluation and assessment of pertinent existing information sources; formulation of the hypotheses; selection of base maps; development of the classification systems and data collecting techniques; and if sampling procedures are to be employed, selection of the sampling system (see Figure 6-3). The planning phase takes place in the office for the most part and requires considerable thought and decision-making. It is the policy-making part of the research project.

The products resulting from the planning stage of the research problem need to be tested under actual field conditions. The reconnaissance serves as a pilot study to determine if the classification systems and data collecting techniques are viable as developed or need modification. The hypotheses may be reworded and refined and, if sampling is necessary, the sample areas may be selected or confirmed at this time. Lastly, the community or research area should be prepared for the forthcoming field research. However, if the project requires input from the community in the development of the research problem, this step needs to be accomplished during the planning phase.

Assuming that the planning and pilot study phases were well structured and have been completed successfully, the actual field work should proceed without undue difficulties. The field data collecting phase normally takes the larger part of the total time allocated to the research problem, and although time-consuming and routine, the field work essentially puts into operation what has been developed during the planning and pilot study phases. When all the field data have been obtained, the task of tabulation and compilation must be undertaken. What methods of analysis and presentation should be employed depends upon the objectives of the problem. Field research does not differ from other types of research in this respect. The analysis of the data may be relatively simple and straightforward or highly sophisticated, again depending upon the magnitude and purpose of the research study. The conclusion phase or the summary of the research findings essentially is the description of the major points of the analysis. The answer or partial answer to the problem statement and the confirmation or denial of the hypotheses are presented as documented by the data collected and analyzed.

The amount or percentage of time to be allocated to each of the phases described above will, of course, vary greatly from one field research problem to another. In most cases, the planning and field data collection phases will consume the greater amount of time. If the size of the research area is large and the project takes months, the field work may take a very high percentage of the total time devoted to the project. In all cases, a carefully devised work/time plan is essential in order to organize the efforts and resources of researchers and associated personnel efficiently.

## Communications

In research problems of large magnitude, communication and information exchange within the project infrastructure are essential. Frequently the field research team may be spread out in widely separated geographical locations and may come together only rarely, or even just for the field data gathering portion of the project. Full communication among researchers in charge of the various subject matter areas of a project is absolutely necessary for proper coordination during the planning, evaluation, and steps requiring uniform philosphical bases or joint efforts. It is of paramount importance that all researchers charged with managerial responsibility for a portion of the project fully understand the entire research design and the role played by their counterparts. Their efforts must dovetail smoothly with work done by other principal researchers.

Frequently the research problems are highly complex and require specialists in diverse fields. The specialists may not be accustomed to dealing with geographers, geographical viewpoints, or even other specialists in different subject matter areas. Therefore, coordination can be difficult. It may be necessary to schedule weekly or bimonthly meetings of such research specialists in a city of convenience (located closest to all individuals). The entire group should fly there and meet face-to-face. This is, of course, expensive; however, in large and complex projects, it is frequently the cheapest method of coordination. It can be much more expensive to unsnarl the mistakes and misunderstandings and duplication of effort that are associated with miscommunication.

## FIELD WORK PLANNING PROCEDURES

Careful planning is necessary prior to beginning the field data gathering process itself. This principal node, shown in Figure 6-2, is a result of eight important input processes:

1. Sample design;
2. Social survey instrument design;
3. Classification unit design;
4. Field reconnaissance;
5. Philosophical viewpoint for the study;
6. Acquisition of field-related materials and equipment;
7. Field worker training; and
8. Community preparation and public relations.

The first three—sample design, social survey instrument design, and classification design—have been discussed in previous chapters. The principal requirement of these processes is that the procedures work in the field situation, are cost effective, and provide the necessary information.

In all cases, it is necessary to build in a feedback loop, illustrated for each case in Figure 6-2, to "proof" and evaluate the procedures before using them in the field. This proofing procedure may require varying degrees of field reconnaissance, including reconnaissance sampling tests or other field experiments that might be required to prepare adequately for the final data gathering experience.

### Acquistion of Field-Related Materials

At this time, it is necessary to obtain all field equipment, aerial photography, and other required materials and test their effectiveness. If air photography is ordered or hired flown, the research team should be prepared for some delays due to normal lag time for acquiring such materials.

### Field Worker Training

Also at this time, each field person should be adequately trained, both generally and for his or her specific task. For example, if estimation techniques are to be employed, enough experimentation time must be allowed so that all members of the survey team can make comparable estimates and/or evaluation of study area phenomena.

### Community Preparation and Public Relations

An important aspect of field work that is frequently overlooked is public relations. To insure full cooperation of the respondents and local residents, the researcher must inform the public of the impending field work. Mapping teams in rural areas may be suspected of poaching, or highly distrusted if local residents believe their activities might raise taxes, bring in a new highway, or make some undesirable change. Local newspapers, radio stations, and community organizations should be contacted. Most will be eager to inform the local people of the nature and duration of the field project. At a minimum, local law enforcement officials should be informed. The names of researchers, vehicle license numbers, the nature of the project, dates, times, and locations should be among the information submitted.

## FIELD WORK AND RELATED ACTIVITIES

With the careful planning and preparation described above, the field data gathering should proceed smoothly. Throughout this phase, quality control and monitoring of field generated products should be made continually. This will insure consistency and data comparability. All information acquired in this step plus the previously gathered secondary information constitute the project data pool.

Following the field work, each person should be thoroughly debriefed. Personal experiences, observations, unsolicited information, and a general project perspective will be gained. All information of this type will be potentially valuable to those persons engaged in data analysis and report preparation.

## ANALYSIS AND SYNTHESIS

The next step is the preparation of the data for analysis or synthesis according to the philosophic framework of the project. The types of techniques employed will vary from project to project. Some will require elaborate coding and preparation for statistical and other analyses by digital computers while others will use less sophisticated methods, such as maps and hand-processed graphs. In any event, a careful checking and proofing of data as it is coded or mapped will reduce mistakes which may easily affect the data reliability. Such proofing occurs not just at the coding phase but also during the analysis and/or synthesis phase as well.

Finally, the results are determined and appropriate cartographic displays and reports are made. Either the question will be answered and a decision may be made based on the result, or the data will not answer the question. If the latter occurs, a possible rehypothesis and additional field work may be necessary. In most cases, however, the project results will be sufficient to allow administrative decisions to be made based on the field investigation and data reduction and analysis.

The methods used to analyze and present field data do not differ much from those employed in other types of research. The only significant difference between a field research problem and other research problems is that the field research study utilizes primary or raw data obtained from observations in the field rather than existing published data. The same care must be exercised to select the most appropriate cartographic and statistical analytical devices. The advantages, limitations, and restrictions of dot, isoline, ratio, or flow maps must be weighed carefully, as well as the values assigned to dots, lines, etc. The statistical devices used may be highly sophisticated or relatively simple depending upon the problems and the philosophy and background of the researcher. It is assumed that the researcher will apply knowledge acquired in other academic courses or experiences to the analysis and presentation of the field data.

## THE IMPORTANCE OF THE RESEARCH DESIGN

No two field research problems are precisely alike. All field research problems must have a research design, but the complexity and intricacy of the design will vary greatly from one problem to another. The research design is the structure or framework for the entire field research problem. It is the blueprint of the project to be undertaken and must be compatible with the needs of the specific problem in terms of the time, capabilities, and interests of the researcher, and the final results desired. It must be realistic and not overly complex or overly generalized. The creation of the research design is the major part of the field research problem and, if it is done soundly, the components of the research itself will mesh together and the goal will be accomplished with accuracy and efficiency.

## PROJECT PLANNING AND PROPOSAL WRITING

The expense of field work is often greater than the individual field researcher can bear and, at some time or another he will probably need to acquire extra funds for a portion, if not all, of a field research project. Because of this proposal writing has become a necessary part of research in this country in recent years. Since requests for funding greatly exceed the monies normally available, the researcher must put considerable thought and effort into writing a proposal if it is to be considered seriously by the funding agencies.

Writing a proposal not only helps the researcher obtain needed funds but it also forces him to think through the proposed project from beginning to end and organize the venture in the most effective way. What is to be accomplished, how it is to be accomplished (methodologies and techniques), why it is important, what the completed project's impact will be, and who will be affected are all questions that must be answered clearly. Writing the proposal itself is a major and valuable diagnostic tool in the organization and justification of the field research work.

Some funding groups provide the proposal writer with guidelines and directives for preparing the proposal, and others do not. However, in all cases, several major components make up a sound and comprehensive proposal, and normally these components are in a specific order. The emphasis placed on each component may vary depending upon the nature of a specific proposal, and in some cases two or more elements may be combined depending upon the agency's specific guidelines. The normal components and steps are:

I. Preliminary investigation
II. A. Covering letter
   B. Proposal summary

C. Budget summary
D. Proposal introduction
E. Substantive statement (problem statement)
F. Project objectives
G. Project design
H. Personnel commitment
I. Qualifications of personnel
J. Assessment plan
K. Potential impact
L. Budget

The substance of the proposal is normally included in items E-L above. Items A-D are considered preliminary or supplementary statements or activities supporting the main body of the proposal.

## Preliminary Investigation

The first step in preparing any proposal is identifying the agencies that appear to have an interest in proposed field work. It is a waste of time and effort to prepare and submit a proposal to a group whose objectives or scope is not compatible with the proposed project. The researcher must make a detailed analysis of announcements, brochures, annual reports, correspondence, and similar materials to determine the agencies that will consider the proposed research, institution, discipline, and re- searchers involved. There is a great variety among funding groups as to what re- search and what institutional structures are eligible or ineligible for consideration. Then too, a funding source's fiscal and temporal limitations may not be practicable for a given field project. An initial investigation will determine what agencies, from the standpoint of *their* objectives and magnitude, are potential sources of funding. Out of several hundred funding groups perhaps only a few will fit the objectives and needs of the proposed field project.

## Covering Letter

Proposals are ordinarily accompanied by a covering letter, which is essentially a statement of intent and commitment. It assures the funding group that the research- ers are committed to the project, and that their superiors and their institution's authorities or administrative officers are aware of the proposed project, endorse it, and will support it. Briefly the letter should describe the proposal, state the amount of funds requested, specify the dates the project would start and end, and name the proposed director and principal members of the team. Phone numbers of the proposed director or persons in authority should be included should the funding agency have any questions. Should any key personnel be out of reach the dates they cannot be reached should be noted.

## Proposal Summary

Normally an abstract appears at the beginning of the proposal. After the final draft of the proposal has been written, however, the highlights can be more readily summa- rized. The summary is the first part of the proposal to be read, and it may be the only part carefully scrutinized by all readers. Further, it may serve as the official statement for public distribution, news releases, and the like at a later date. For these reasons, it should be written in a clear and concise manner so a person not familiar with the discipline or the area of research can understand the nature of the project and the need for funds to support the work.

## Budget Summary

In many cases, the funding source provides instructions and forms for preparing the budget. The usual format includes personnel costs (wages, salaries, consultant fees, etc.), nonpersonnel costs (supplies, equipment, computer rental, etc.), and indirect costs (charges for the use of facilities and upkeep). Indirect costs are normally an agreed upon percentage of the project's direct costs. The budget summary need not be concerned with listing details or justifying expenditures. The budget proper at the end of the overall proposal usually contains those items. The budget summary along with the summary of the proposal present a basic overview of the entire proposal; together they enable the funding source to determine quickly the general nature and cost of the proposed project.

## Proposal Introduction

The introduction acquaints the readers with the history, objectives, and philosophy of the department or institution requesting funds. It establishes the responsibility and suitability of the organization that will manage the proposed field project. The proposal writer should list the institution's credentials and describe the facilities available for carrying out the project as well as similar or related projects currently underway or recently completed. Other information supporting the institution's credibility should be included at this point.

## Substantive Statement

The substantive statement is the heart of the proposal and is usually the longest and most detailed component. It describes in detail what the project is and why it needs to be funded. The researcher should include a documented review of the pertinent literature as well as a list of current and past activities related to the proposed project; statistical data and authoritative statements may enhance this review. This section explains the project's raison d'etre and needs to be prepared with great care; other components of the proposal merely provide supporting evidence and details.

## Project Objectives

A clear statement of how the project will contribute information and knowledge to others, the project objectives should indicate the anticipated outcomes important to a given set of concepts, theories, or bodies of knowledge. What groups will benefit most from the results of the project, and how they will benefit should be described. Again, statistical information and authoritative statements should be included if at all possible.

## Project Design

The project design is a detailed description of how the proposed project will be carried out from start to completion. Methodologies and techniques to be employed should be discussed in detail, and work schedules, timetables, deadlines, and major divisions for the overall project activities should accompany that description. Graphic illustrations like flowcharts, tables, graphs, and schedules can emphasize the highpoints. What tasks must be accomplished, who will perform those tasks, and, how and when those tasks must be completed are details to be described in the project design.

## Personnel Commitment

The cover letter contained a brief statement committing the researchers and the institution to the proposed project, but the proposal writer should elaborate on who will be involved, and when precisely they will be working on the project. If any project personnel are to be part-time, it is necessary to spell out how much of their time will be devoted to the project and how much time will be spent on other specified duties. It is important to assure the funding group that the project personnel and the institution as a whole are most interested in the proposed work, and, if funded, will devote their efforts and facilities enthusiastically to complete the proposed project.

## Qualifications of Personnel

Each of the professionals having responsibilities in the proposed project should be described with a brief outline of education, training, and experience directly related to the proposed project's work, and a detailed resume or curriculum vitae (to be included in an appendix). This statement assures the funding group that qualified people will be involved in the project's work.

## Assessment Plan

Many times a mechanism to evaluate the success of the project is necessary. An assessment plan may also serve as a device to obtain data for improving the project and perhaps further refining it. The plan should be a clear statement and describe how the evaluation will be done, and who will do it. The way criteria for evaluating the project are defined and the manner in which the information to assess its success is acquired and processed vary greatly; both are critical components of the overall assessment process. Each project is unique and the assessment procedure must often be tailored to fit the specific project.

## Potential Impact

All completed field projects have an impact on some group; it may be local, regional, or even national in magnitude. The potential impact should be described as specifically as possible. This will further support statements concerning the importance of the project and the need for financial support. In addition the results of the project may suggest that the project itself be continued or spin-off projects be developed.

These possibilities should be discussed in detail to provide the funding group with a long range plan if the project proves to be especially successful and productive.

### Budget

The budget summary included a brief statement listing the major areas of proposed expenditures but not detailed information or justifications. In contrast this section should be a detailed account of specific anticipated expenditures for the project. It is important to prepare the budget with great care, closely adhering to the funding group's directives, guidelines, and limitations. In every case, the budget must be specific, fully justify all expenditures, and be fully compatible with the rules and regulations of the funding group and the researcher's institution.

In the last analysis, a good proposal will state clearly what the researcher intends to do, why it is important, how he intends to do it, the qualifications to the personnel who will do the work, the facilities and equipment to be used, the potential impacts, and the precise cost.* This is the information proposal readers or reviewers look for and evaluate.

## SUGGESTED REFERENCES

ABLER, R., J. S. ADAMS, and P. GOULD, *Spatial Organizations.* Englewood Cliffs, New Jersey: Prentice-Hall, 1971.

ACKERMAN, E. A., *Geography as a Fundamental Research Discipline,* Department of Geography Research Paper No. 53. Chicago: University of Chicago Press, 1958.

AMEDEO, D. and R. G. GOLLEGE, *An Introduction to Scientific Reasoning in Geography. New York: John Wiley & Sons,* 1975.

BUNGE, W., *Theoretical Geography,* Lund Studies in Geography. Lund, Sweden: The Royal University of Lund, C.W.K. Gleerup, 1966.

CHORLEY, R. J. and P. HAGGETT, eds., *Models in Geography.* Toronto, Ontario: Methuen Publications, 1967.

COOKE, R. U. and J. H. JOHNSON, eds., *Trends in Geography: An Introductory Survey.* London: Pergamon Press, 1969.

DURRENBERGER, R. W., *Geographical Research and Writing.* New York: Thomas. Y. Crowell Co., 1971.

FITZGERARD, B. P., *Science in Geography: Developments in Geographical Method.* London: Oxford University Press, 1974.

HAGGETT, PETER, *Geography: A Modern Synthesis.* New York: Harper & Row, Publishers, 1972.

HARING, L. L. and J. F. LOUNSBURY, *Introduction to Scientific Geographic Research.* Dubuque, Iowa: William C. Brown Co., Publishers, 1975.

HARVEY, D. W., *Explanation in Geography.* New York: The Natural History Press, 1970.

INHABER, H., *Environmental Indices.* New York: John Wiley & Sons, 1976.

McCULLAGH, PATRICK, *Science in Geography: Data Use and Interpretation.* London: Oxford University Press, 1974.

TAAFFE, E. J., ed., *Geography.* Englewood Cliffs, New Jersey: Prentice-Hall, 1969.

---

* An excellent source describing proposal writing is N.J. Kiritz's *"Program Planning and Proposal Writing,"* in *The Grantmanship Center News,* (May/June, 1979).

# Appendix A
## Selected Examples of Field Exercises

---

The main objective of the geographic field course is to develop the student's observational skills and perspectives. The primary objective of the comprehensive field course is to expose the student to a wide range of field methods and techniques. These experiences will provide a pool of knowledge from which he or she can select, refine, and apply approaches to meet the needs of most future field problems. One comprehensive field course will not make the student a master of all phases of data gathering, but he or she will be knowledgeable in the sense of knowing how and when to employ a given technique.

The exercises included in this appendix are hypothetical (although modified from actual exercises used by the authors for several years) and have been selected to represent a range of field problems. They are only suggested structures and approaches and may be altered with ease to meet the needs of a particular course and area.

For the introductory comprehensive field course, field exercises should be designed to represent various stages of difficulty and complexity. They may be classified into introductory exercises, intermediate level, and advanced field problems.

The introductory field exercises should be utilized to introduce the inexperienced student to field study and, therefore, should be highly structured, allowing the student little freedom of choice. They should only be concerned with measurement or data collecting and require little or no analysis. The emphasis should be on acquiring skills in a limited number of techniques. Exacting instructions should be provided and the matrix of the field work precisely defined. The exercises should be as short as possible, but designed so that they can be completed in the time allowed. The basic principles of a given technique can be grasped in a period of a few hours if the exercise is designed properly. An exercise taking a few days would no doubt increase the skill and efficiency of the student, but little additional conceptual knowledge would be acquired (see Tables A-1, A-2, A-3, and A-4).

Exercises at the intermediate level are designed for students who have had some field experience. They are structured to include specific instructions, but require some ingenuity and analysis. Often several types of data collecting techniques are required, and the field exercise may relate to an established methodology. Intermediate exercises should take from one to three days and require techniques other than data collecting and measurements (see Tables A-5, A-6, A-7, A-8, and A-9).

The advanced field problems are designed for the student who has acquired the knowledge and skills of geographic field research. The exercises should provide some direction, but their purpose is to force the student to develop and complete field research problems by applying whatever methods and techniques are the most effective. They require the student to think through the problem and plan the research design from start to finish (see Tables A-10 and A-11). They may be considered the capstone of the student's field training.

These hypothetical illustrations of field exercises are but a few selected examples. Similar exercises concerned with historical geography, suburbanization, microclimates, etc., can be developed at various stages of difficulty to focus on a specific area. Remember, however, that the primary objective of all field exercises, regardless of level, is to train students in field data collecting techniques and analysis. The exercises provide the student the opportunity to apply theory, academic training, and experience to the real-world situations. Although training and acquiring experience are the goals, every effort should be made to make the exercises as interesting and stimulating as possible.

## SUGGESTED REFERENCES

BOARD, C., "Field Work in Geography, with Particular Emphasis on the Role of Land-Use Survey," *Frontiers in Geographical Teaching*, edited by R. J. Chorley and P. Haggett, 186–214. London: Methuen Publications, 1965.

CROSS, M. F. and P. A. DANIEL, *Fieldwork for Geography Classes*. New York: McGraw-Hill Book Co., 1968.

EVERSON, J. A., "Some Aspects of Teaching Geography Through Fieldwork," *Geography*, Vol. 54 (Part I), No. 242, 64–73, January 1969.

FRIBERG, J. C., *Fieldwork Techniques: A Revised Research Bibliography for the Fieldworker and Reference Guide for Classroom Studies*, Discussion Paper No. 14, Department of Geography, Syracuse University, 1976.

HART, J. F., *Field Training in Geography*, Commission on College Geography Technical Paper No. 1, Association of American Geographers, 1968.

HUNT, A. J., "Land-Use Survey as a Training Project," *Geography*, Vol. 38, 277–286, 1953.

HUTCHINGS, G. E., "Geographical Field Training," *Geography*, Vol. 47, 1–14, 1962.

WHEELER, K. S., ed., *Geography in the Field.* London: Anthony Blond, 1970.

---

## TABLE A-1.

*Plane-Table Exercise.*

---

*Operational Procedure*

Exercise to be done in teams of two or three; length of time 3–5 hours.

*Objectives*

To learn how to construct a base map of a small area. Such maps can then be used to map land use, settlement features, or any landscape phenomenon. The student will become acquainted with cadastral measurements, the use of instruments, mapping techniques of intersection and triangulation, establishment of reference points and lines of traverse, and the important concept of scale. Plane-table mapping as such is used in geographic field work only in rare instances, but associated cadastral measurements are basic to several types of microscale problems.

*Problem*

This particular exercise is to construct a topographic map of a small section north of Townsville. The map should be made at the scale of 1:600 (1 inch = 50 feet). The basic principle is to determine accurately (1) direction, (2) distance, and (3) differences in elevation* of one point to others. These data, precisely recorded on the plane-table map, provide a series of known reference points which serve as the basic framework of the map. The mapping area is between Oak Road, Maple Lane, and Beaver Creek (see Figure A-1).

   The finished product will be a map that includes a closed traverse along the Oak Road, Maple Lane, and Beaver Creek; contours at 5-foot intervals;* and the location and identification of all buildings or structures within the map boundaries. Scale and direction should be indicated as well as an appropriate legend.

*Equipment Needed*

Telescopic alidade* or open-sight alidade.
Philadelphia rod.*
Plane table and tripod.
Range or chaining pins.
Rule or scale.
Straight pins, paper, and pencil.

---

*Not pertinent if an open-sight alidade is utilized.

---

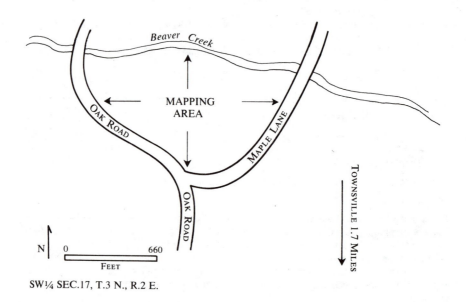

N    0           660
FEET

SW¼ SEC.17, T.3 N., R.2 E.

**Figure A-1.**

Locational Map for Hypothetical Plane-Table Exercise.

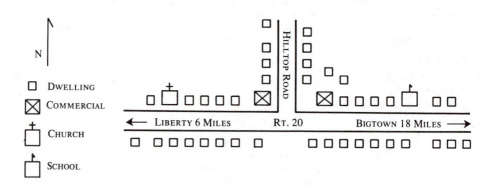

**Figure A-2A.**

Example of a Base Map for a Hypothetical Small Community Exercise.

## TABLE A-2.

*Small Community Exercise.*

*Operational Procedure*

Exercises to be done in teams of two; length of time 1 day.

*Objectives*

To learn how to obtain information in the field employing the interviewing technique of data collection. When tabulated, the data will reveal, to some degree, the function of a small urban community and may be compared with data from other communities of similar size or larger urban agglomerations.

*Problem*

This particular exercise is to acquire selected data concerning the population and function of a small community (hamlet or village) of about 50 to 150 total population. The information to be obtained is: (1) total population of the community; (2) generalized age groupings; and (3) place of work of the gainfully employed. Specifically, a team of two students will collect data in each of the small communities to be surveyed. A base map, not necessarily drawn to scale, must be used to show the precise number of dwellings in the continuous built-up community; the location of the dwellings may be approximate (see Figure A-2A). A questionnaire for each of the dwellings must be employed (see Figure A-2B) and the resulting data tabulated.

Frequently, small communities are essentially residential in function. They serve as a place of residence for persons working in other localities. In addition, the population age groupings are noticeably different when compared with city or state or national age profiles.

Tabulate the data collected to include (1) total population of Village *X*; (2) age groupings; (3) total number and percentage of the gainfully employed working outside of the community; and (4) number and percentage of the gainfully employed working in Bigtown and in Liberty.

*Equipment Needed*

Questionnaire.
Base map showing number of dwellings within the community.

---

B. EXAMPLE OF A QUESTIONNAIRE.

1. How many people live at this residence? _____ Total number
2. Where do the gainfully employed work? Number
   - Village *X* _____
   - Bigtown _____
   - Liberty _____
   - Other _____
3. What are the ages of the people living at this residence?      Number
   - Under 18 years of age _____
   - 18–35 years of age _____
   - 36–60 years of age _____
   - Over 60 years of age _____

---

## Figure A-2B.

Example of a Questionnaire for a Hypothetical Small Community Exercise.

## TABLE A-3.

*Sampling Exercise.*

*Operational Procedure*

Exercise to be done in teams of two, with an even number of teams; length of time 1 to 1½ days.

*Objectives*

To provide experience in field sampling as well as to allow a comparison of the results of several spatial sampling procedures against the characteristics of a known population. Random sampling, stratified sampling, and a variation of stratified random sampling will be used.

*Problem*

Within the spatial matrix of the study area, which of the three sampling techniques produces data that best represents the known population?

*Procedure*

Three easily observed physical characteristics of the residences of Centerville will be enumerated—type of yard fence, type of roof, and the presence or absence of ventilator "turbines" on the roof. A standardized mapping notation code is to be employed. The code consists of three digits with the appropriate number selected from the list of possibilities.

Example:

251

2 = Reinforced concrete fence
5 = Tile roof
1 = Ventilator turbine

Fence Type (dominant type in linear feet)
       1 Cement Block
       2 Reinforced concrete
       3 Brick
       4 Wood (solid)
       5 Wood (rail)
       6 Wood (picket)
       7 Wire (horizontal strands only)
       8 Wire (woven)
       9 Other
       0 None

Roof Type (dominant type in square feet)
       1 Composition shingles
       2 Shake shingles (wood)
       3 Roll composition
       4 Tar paper
       5 Tile
       6 Rock or gravel (hot mopped tar)
       7 Tin
       8 Other
       9 None

## TABLE A-3. (cont.)

*Sampling Exercises.*

Turbines

    1 Yes
    2 No

Only houses or duplexes are to be enumerated—no apartments or non residential buildings.

Teams will be paired and assigned adjacent mapping areas of nine square blocks each to facilitate their working on both areas (one area for 100 percent documentation of existing features, the other area for the sampling trials). Base map information and final results will be traded between the teams.

Population documentation. Teams will begin by mapping the rough location of each house and listing the fence/roof/ventilator characteristics for each. Teams will produce maps on graph paper and use mapping/notation coding for each house. Each team will provide house locations to the other team assigned to that area. (For instance team 1 is assigned area 1 to document characteristics, and area 2 to sample. Team 2 is assigned area 2 to document characteristics and area 1 to sample.) Each team will provide a complete coded map with the total population documentation to the other team when it has completed the sampling and is ready to evaluate results.

Sampling. After obtaining the house location base map from the other team, the sampling phase will begin.

    Make two copies of this map. On one copy lay out an X, Y coordinate set using inches and tenths along the bottom and left hand margins of the page. The lower left hand corner is the origin. Then randomly allocate x-y coordinates for a number of points equal to 20 percent of the houses. Mark the house closest to each point for later sampling. If the closest house has already been selected, use another set of random coordinate pairs. This constitutes the "random sample."

    On the second map construct a regular grid network of vertical and horizontal lines so that the intersections of the grid lines number 20 percent of the houses. Select the house nearest to each grid intersection. If a house has already been selected choose the next closest. This constitutes the "stratified sample."

    Using the third map, group all of the houses into clusters of 5 (i.e., circle closest groups of 5). Then use the random numbers table to select one of the five for the sample in that area (20 percent of each cluster). This constitutes the "stratified random sample."

    Teams may now visit the designated houses and gather information. The collected data will provide the basis of a tabular report showing the percentage of each phenomena under each of the three sampling schemes and also showing the 100 percent enumeration information. A brief written report on the results concludes the project.

*Equipment needed*

Random numbers table.
Graph paper.
Pencils and ruler.

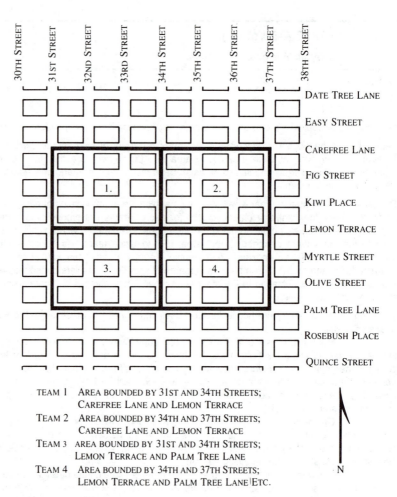

TEAM 1   AREA BOUNDED BY 31ST AND 34TH STREETS;
         CAREFREE LANE AND LEMON TERRACE
TEAM 2   AREA BOUNDED BY 34TH AND 37TH STREETS;
         CAREFREE LANE AND LEMON TERRACE
TEAM 3   AREA BOUNDED BY 31ST AND 34TH STREETS;
         LEMON TERRACE AND PALM TREE LANE
TEAM 4   AREA BOUNDED BY 34TH AND 37TH STREETS;
         LEMON TERRACE AND PALM TREE LANE|ETC.

## Figure A-3.

Locational Map for Hypothetical Sampling Exercise.

**TABLE A-4.**

*Stream Channel (Arroyo) Exercise.*

*Operational Procedure*

Exercise to be done in teams of three or four; length of time 1 day.

*Objectives*

To improve skills in (1) field observation and the decision-making process related to measurement of physical parameters on the real landscape where existing conditions do not always fit those expected; (2) formulation of a working definition of physical feature limits based on observations within a given spatial matrix; (3) use of topographic maps to guide data acquisition; (4) methods of analysis of the raw data collected; and (5) comparison of results with the existing literature for stream channel geometry and relationships.

*Problem*

Moving from one point to another along an arroyo or dry stream channel, the measureable characteristics of the channel can be seen to change. What kinds of relationships occur among the length, width, depth, and bed angle as each of these measurements vary?

*Hypotheses*

1. The depth of channel decreases at a constant rate going upstream.
2. The width of channel decreases at a constant rate as the distance upstream increases.
3. The elevation of bed increases at an increasing rate going upstream.
4. The width of channel increases as the depth of channel increases.
5. As the elevation of bed decreases, the depth of channel decreases at a constant rate.
6. As the stream channel widens, the elevation of bed decreases at a decreasing rate.

*Procedure*

One arroyo will be assigned to each group based on accessibility and length. Starting at the specified initial point, measurements are to be taken every 50 feet upstream along the channel as far as it is practical to do so (see Figure A-5). At each 50-foot station, the following information is to be obtained: (1) bank-to-bank width of channel; (2) depth of channel; (3) elevation of stream bed.

Before collecting the data, each group must walk along its arroyo and observe the degree of definition of the stream channel itself since it is a component within the larger stream valley. Following periods of rainfall, channel definition will be somewhat straightforward; however, with increasing time since the last precipitation which caused water to flow in the channel, the boundaries will become less distinct and decisions need to be made with regard to measurement procedures for consistent data acquisition. (Following the exercise, a class discussion of the operational definition and measurement difficulties encountered will occur.)

*Method of Analysis*

The relationships and rates of change will be found by preparing scattergrams of all possible pairs of data variables, including the distance upstream from the initial point of departure. Six graphs will be constructed and a line *estimated* through the data. If the observed rate of change is

155

## TABLE A-4. (cont.)

*Stream Channel (Arroyo) Exercise.*

constant, a straight line will fit the data cluster; if it does not, then try a curved line (see Figure A-4).

Write a short statement to identify the fate of the hypotheses and to summarize your findings and relevant experiences.

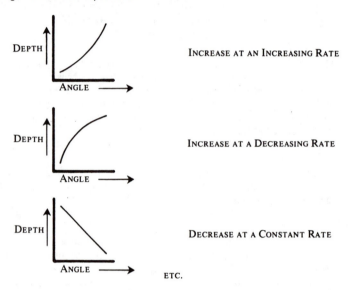

INCREASE AT AN INCREASING RATE

INCREASE AT A DECREASING RATE

DECREASE AT A CONSTANT RATE

ETC.

### Figure A-4.

Graphs of Possible Relationships.

*Equipment Needed*

Chain.
Cloth tape.
Abney level or equivalent (protractor and plumb line).
Field notebook.

## SUGGESTED REFERENCES

LEOPOLD., L. B. and T. MADDOCK, "The Hydraulic Geometry of Stream Channels and Some Physiographic Implications," U.S. Geological Survey Professional Paper No. 252.

LEOPOLD, L. B. and J. P. MILLER, "Ephemeral Streams—Hydraulic Factors and Their Relation to the Drainage Net," U.S. Geological Survey Professional Paper 282-A.

ROBERTS, M. C ., "The Hydraulic Geometry of the Middle and Upper Portions of Mosca Creek, Sangre De Cristo Range, Colorado," *Proceedings of the Oklahoma Academy of Science,* Vol. 46, pp. 205–212, 1966.

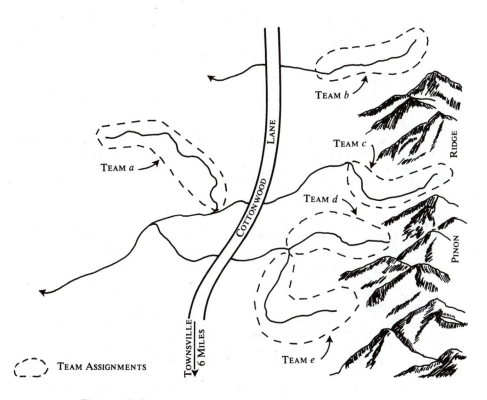

TEAM b

TEAM c

TEAM a

TEAM d

RIDGE

PINON

TOWNSVILLE
6 MILES

COTTONWOOD LANE

TEAM ASSIGNMENTS

TEAM e

**Figure A-5.**

Locational Map for Hypothetical Stream Channel Exercise.

## TABLE A-5.

*Central Business District (CBD) Exercise.*

*Operational Procedure*

Exercise to be done individually or in teams of two; length of time 1½ to 2 days.

*Objectives*

To improve skills in (1) identification and classification of urban features; (2) mapping of techniques associated with urban land use and dimensional mapping; (3) methods of analysis of the raw data collected; and (4) application of established models to the specific field exercise.

*Problem*

To analyze the functions and structures of the central business district (CBD) of Metropolitanville (100,000-1,000,000 + population), and to devise methods of delimitation. In the great majority of cities, a viable CBD differs from other commercial areas in that within the CBD,

1. A great variety of commercial enterprises is in evidence;
2. The maximum concentration of commercial land uses is found;
3. The highest land values and real estate taxes of the greater metropolitan area exist;
4. All commercial uses must realize a high financial return per unit of space (the greatest volume of business per square foot);
5. Multistoried buildings and vertical zonation predominate;
6. The highest traffic count, both vehicle and pedestrian, of the greater metropolitan area is found;
7. The structure and profile of commercial uses reflect accessibility to the total metropolitan area; and
8. Buildings and stores are well maintained. Normally, the CBD is surrounded by a blight zone or area of deterioration.

*Hypotheses*

1. The CBD of Metropolitanville reflects the characteristics stated above.
2. The CBD of Metropolitanville may be delimited on the basis of land use, vertical zonation, and general appearance of the buildings and individual stores.

*Method of Analysis*

In a two-day period, the class as a whole will delimit and analyze the CBD of Metropolitanville. Data will be collected, assembled in a master package, and analyzed collectively. The following two methods of analysis and delimitation will be employed:

1. *The Transect Method.* Each individual or team of two persons will map a linear traverse eight blocks in length (from Elm Street to Cedar Street) through the CBD (see Figure A-6). In width, the traverse will include the land use of the street frontage *only* on both sides of the street. Mapping may be done on graph paper at the scale of 1 inch = 100 feet (see Figure A-7). The data shown on the plat map must be drawn on the graph paper at the proper scale. Multifeature mapping is necessary to include (A) land use, (B) height of buildings, and (C) quality or upkeep of buildings and structures. "Breaks in continuity" must be recorded accurately.

158

### TABLE A-5. (cont.)

*Central Business District (CBD) Exercise.*

(A) Land Use: Studies of commercial land uses reveal that specific commercial uses are commonly found within the CBD and that other commercial uses are rarely associated with the CBD, as noted in the following list.

| *Normally Not CBD Uses* | *Common CBD Uses* |
|---|---|
| Manufacturing plants | Banks, savings and loan associations |
| Permanent residences (single-family dwellings, rooming houses, apartments) | Professional offices (law, medical, and insurance offices) |
| Schools, churches, parks, fraternal orders | Department stores |
| Supermarkets, convenience markets, grocery stores | Men's and/or women's clothing and shoe stores |
| Discount and variety stores | Jewelry stores and gift shops |
| Automobile dealerships | Restaurants |
| Service stations and garages | Hotels |
| Open or outdoor parking lots | Theaters and first-run movie houses |
| Warehouse and storage facilities | National and regional corporation offices or headquarters |
| Taverns and bars | |
| Wholesale establishments | Nightclubs and cocktail lounges |
| Second-run movies and burlesque houses | Drugstores |
| Pawnshops | Indoor garage or multistory parking |
| Governmental agencies (city, state, county) | |
| Vacant buildings and vacant lots | |

For each block (street frontage), determine the percentage of land use (use urban land mapping key). If 75 percent of the block is devoted to normal CBD uses, it is a block within the CBD.

(B) *Building Height:* If the street-front buildings within a block average three or more stories in height, the block may be considered within the CBD.

(C) *Quality of Upkeep:* Each individual or team of two will make subjective judgments as to the quality or condition of the buildings and structures. Judgments should include the general state of repair; condition of paint and masonry; cleanness of windows, doors, entrances, etc.; and aesthetic quality of window and other displays and advertising. If the front footage is generally kept up well, the block may be classified within the CBD.

Each transect or traverse is a sample or cross section of the CBD. The compilation of all of the traverses on a master map will indicate the size and areal extent of the CBD.

2. *Total Block Analysis.* Each individual or team of two persons will map the land use of an entire block at the scale of 1 inch = 50 feet (see Figure A-8). The data on the plat map must be translated to the graph paper at the proper scale. The land uses of the ground floor, second floor, and upper floors (three maps total for each block) must be mapped. Calculate the square footage of each function or use of the block mapped and calculate the Central Business Height Index (CBHI) and the Central Business Intensity Index (CBII) using the Murphy-Vance method, as follows:

CBHI = Central (commercial) space/Total ground-floor space

CBII = Central (commercial) space/Total all floor space × 100

---

### TABLE A-5. (cont.)

*Central Business District (CBD) Exercise.*

---

To qualify as a CBD block, a CBHI of 1 or more and a CBII of 50 percent or more are necessary. Each individual or team will transfer the calculated CBHI and CBII to the master map, and the areal extent of the total or partial CBD will be indicated.

*Equipment Needed*

One sets(s) of appropriate plat maps (to be used as the basis of the graph paper base maps and master map for final presentation).
Graph paper.
Urban land-use mapping key.
Pencil.

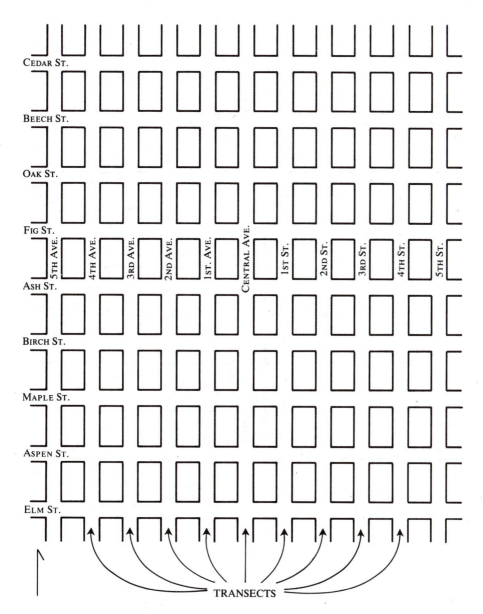

**Figure A-6.**

Metropolitanville Central Business District.

161

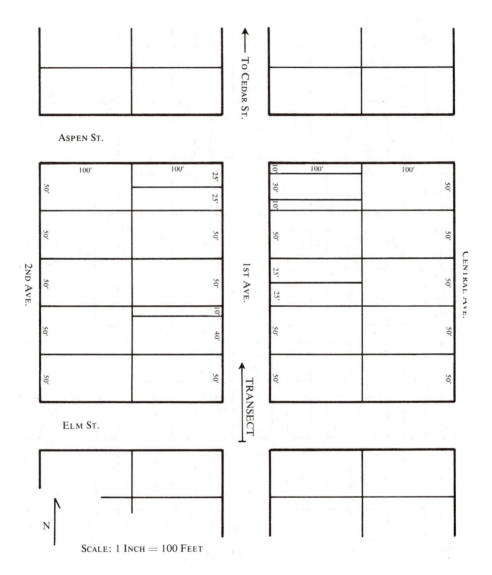

**Figure A-7.**

Sample of Base Map for a Eight-Block Transect or Transverse. To be constructed on graph paper from the data provided on the plat map.

162

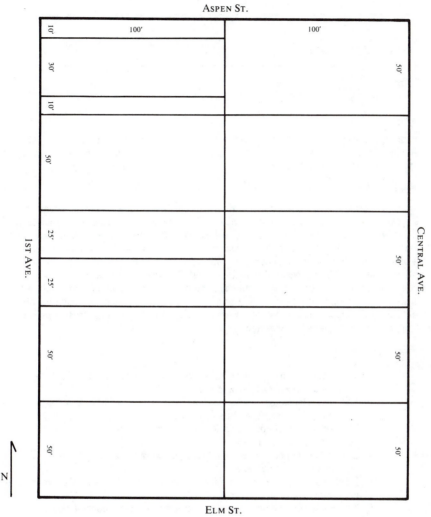

ASPEN ST.

1ST AVE.

CENTRAL AVE.

ELM ST.

N

SCALE: 1 INCH = 50 FEET

## Figure A-8.

Sample of Base Map for Total Block Analysis. To be constructed on graph paper from the data provided on the plat map.

## TABLE A-6.

*Rural Land-Use Exercise.*

*Operational Procedure*

Exercise to be done in teams of two; length of time 2½ to 3 days.

*Objectives*

To improve skills in several data collecting techniques such as rural land-use and fractional-code mapping, and interviewing; and in the analysis of the field data collected. A written report will allow creativity in synthesizing and interpreting the results of the field study.

*Problem*

To analyze a portion of an agricultural and rural environment as to its functions, development, and present trends. Specifically, a two-square-mile area in the vicinity of Citrusville and Avocado will be studied by a team of two (see Figure A-9). The land use, slope, soil types, drainage, erosion, and stoniness of the surface should be mapped for the total two-square-mile area and for each farmstead, rural nonfarm dwelling, industry, etc., interviewed. The analysis of this data will provide the basis of a written report. The major aspects of the report might include the confirmation or denial of the hypotheses; description of the physical setting; analysis of the land-use patterns and the area's economy; present settlement patterns; historical development; and trends and processes active in affecting change. Pertinent maps, tables, and other illustrations should be included.

*General Hypotheses*

1. The elevation of a specific area and subsequent frost hazard is a major determinant in agricultural land use.
2. The slope factor as it influences air and water drainage is a major factor influencing agricultural land-use patterns.
3. The texture, structure, and fertility of specific soil types and phases are major determinants of crop choice.
4. Due to the high cost of land and labor, crops produced are those that realize high-value returns per acre.
5. Land-use changes from predominantly pastoral uses to residential-exotic farming are presently taking place.
6. The high-value and unique agricultural commodities produced enjoy a regional or national market.
7. Labor costs and rising land prices are affecting the types of agricultural commodities produced.
8. Due to the location of the area in relation to the greater Citrusville metropolitan area, many of the farmers are part-time farmers or are semiretired and have additional (nonfarm) sources of income.

*Equipment Needed*

Vertical aerial photo (1 inch = 400 feet).
Rural land-use mapping keys.
Classification of the physical characteristics of the land.
Farm questionnaire.
Slope indicator.
Appropriate topographic quadrangle, USGS 1:24,000 (for orientation purposes only).
County soil survey.

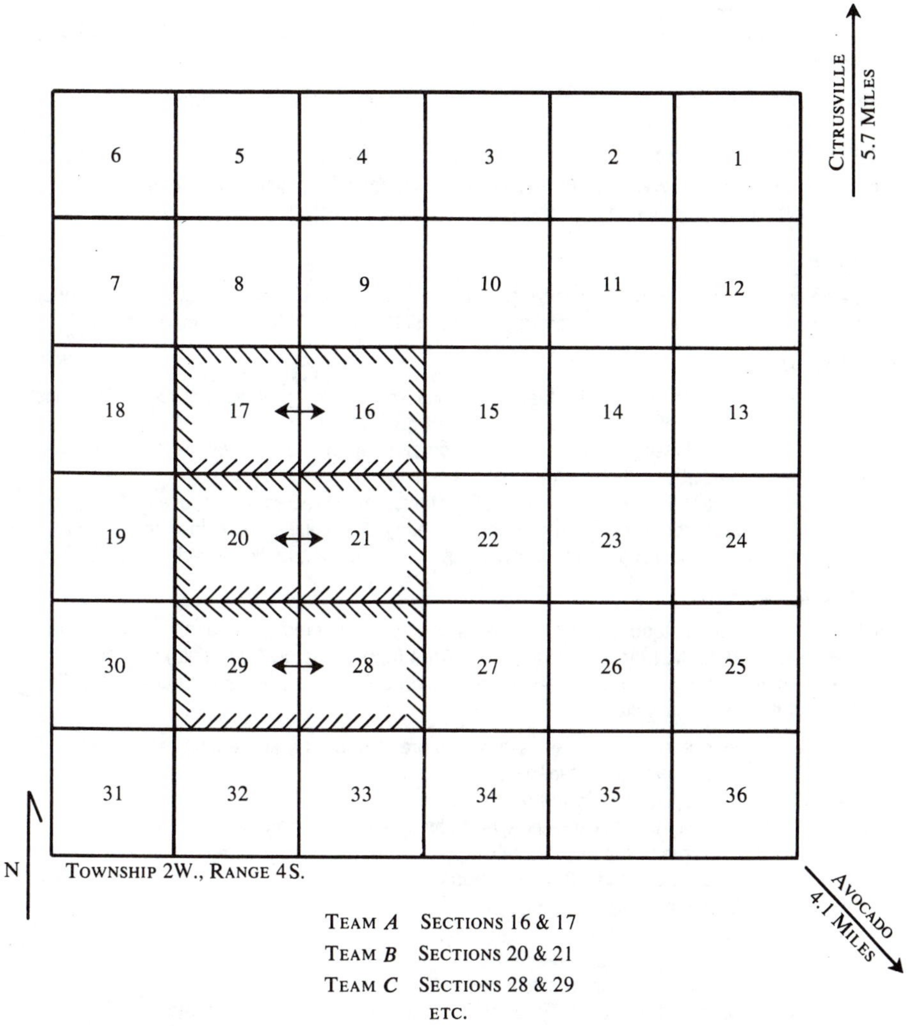

**Figure A-9.**

Locational Map for Hypothetical Rural Land-Use Exercise.

### TABLE A-7.

*Recreational Exercise.*

*Operational Procedure*

Exercise to be done individually or in teams of two; length of time 1 to 1½ days.

*Objectives*

To improve skills in observation; developing a questionnaire; obtaining data employing interviewing techniques; and methods of analysis of the data collected.

*Problem*

To inventory and analyze the morphology, function, and user characteristics of two outdoor recreation facilities in the vicinity of Metropolitanville (100,000-1,000,000+ population).

*General Hypotheses*

1. Frequency of individual user visits are directly proportional to the distance from the facility rather than the activity offered.
2. The drawing power of a recreation site depends on the variety of experiences available rather than any single activity.
3. Due to the "get away from it all" attitude among recreationists, users of the parks tend not to congregate in "local" areas but are dispersed in their facility usage.
4. The most preferred functional level of a park is the "local park."

*Method of Analysis*

In 1 or 1½ days, each student or team of two students will inventory and analyze two outdoor recreation facilities, Lake Pleasant Recreational Area (approximately 300 acres) and Washington Park (approximately 12 acres) in the greater Metropolitanville area (see Figure A-10). The methods of the inventory include:

1. Inventory and map (or sketch) on graph paper the recreation areas showing:
    a. Location of *all* facilities,
    b. State of repair and maintenance.
2. Classify the facility according to its functional hierarchy.
3. Observe users to determine:
    a. Spatial variation of use intensity,
    b. Average age of user,
    c. Type of activity.
4. Interview users to determine:
    a. Distance from residence,
    b. Whether they come there from residence, work, school, etc.,
    c. Method of transportation,
    d. Length of stay,
    e. Frequency of visit,
    f. What activity or facility is most important in drawing people there.

The data collected will provide the basis of a written report, to include a comparative analysis of the two recreational areas and the confirmation or denial of the hypotheses stated. Graphs, maps, tables, photographs, etc., should be included.

*Equipment Needed*

Graph paper.
Questionnaire (designed by the researcher or research team).
Paper and pencil.

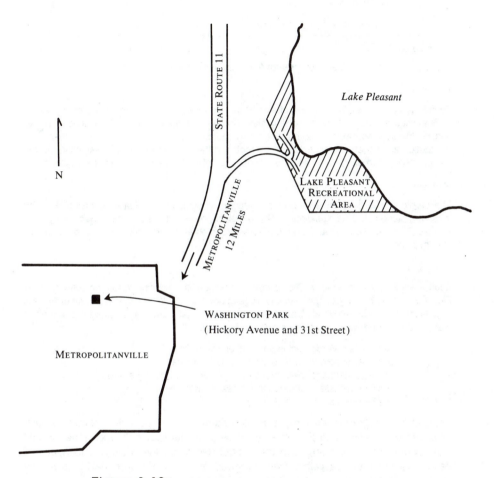

**Figure A-10.**

Locational Map for Hypothetical Recreation Exercise.

## TABLE A-8.

*Vegetation Mapping Exercise.*

*Operational Procedure*

Exercise to be done in teams of three or four; length of time 2 to 3 days.

*Objectives*

To improve skills in (1) observation of significant and measurable characteristics of vegetation communities as they relate to existing classification schemes developed by ecologists, (2) relating the species mixture and physiognomic variability of an actual site to the generalized distribution and vegetation abstractions presented in the literature, (3) spatial sampling, (4) the use of aerial photo base maps and the identification of homogeneous areas of surface cover.

*Problem*

The problem is to produce a map of the vegetation communities that exist on the Rancho Westerno properties. These holdings will hypothetically be developed into low-density ranch-style estates. The map would enable an effective environmental impact assessment to be made for the development.

*Procedure*

Each team is to be assigned a one-square-mile area of the Rancho Westerno property (see Figure A-11). According to the regional vegetation map (A. W. Kuchler's Potential Natural Vegetation of the Conterminous United States), there are at least five vegetation communities that could be found within the study area (see Figure A-12):

> #43—Palo Verde-cactus shrub (Cercidium-Opuntia)
> #42—Creosote bush-bursage (Larrea-Franseria)
> #58—Gramma-tobosa shrubsteppe (Bouteloua-Hilaria-Larrea)
> #44—Creosote bush-tarbush (Larrea-Flourensia)
> #27—Mesquite bosque (Prosopis)

Using the photographs, species lists, site criteria, and vegetative parameters such as cover, dominance, and presence-absence of critical species found in the manual that accompanies the map, each team must decide on the limits and locations of vegetative communities within their assigned areas (see Figure A-13). It will be necessary to prestratify the 1:10,000 scale aerial photography supplied for the study to determine homogeneous surface cover types for field sampling. The type of sampling is left to each team. All intergrade areas and obviously human-modified types should be identified. The final map for compilation of the results is a 1:12,000 enlargement of a 1:24,000 topographic quadrangle. Each team will enter its respective results on this compilation.

*Equipment Needed*

Air photography.
Topographic maps.
Species list.
Soil sampling shovel.
Abney level.
Chain.
Cloth tape.
Notebook.

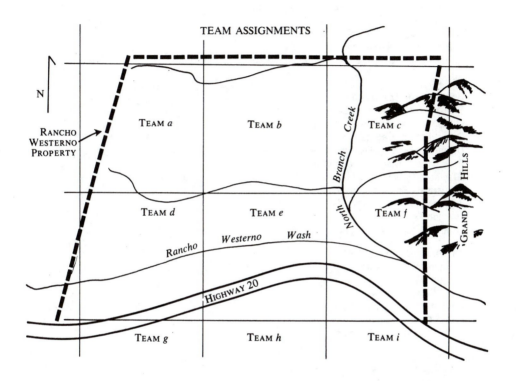

**Figure A-11.**

Locational Map for Hypothetical Vegetation Mapping Exercise.

**Trees**

Blue Paloverde (*Cercidium floridum*) along drainages
Desert Willow (*Salix ssp.*)
Honey Mesquite (*Prosopis juliflora*)
Ironwood (*Olyneya testosa*)
Yellow or Hill Paloverde (*Cercidium microphyllum*) on interflues

**Shrubs**

Bur Sage (*Ambrosia deltoidea*)
Brittlebush (*Encelia farinosa*)
Creosote Bush (*Larrea Tridentata*)
Desert Broom (*Baccharis sarothroides*)
Desert Buckwheat (*Eriogonum ssp.*)
Fairy Duster (*Calliandra eriophylla*)
Graythorn (*Krameria Grayi*)
Groundsel (*Senecio monoensis*)
Jajoba (*Simmondsia chinensis*)
Mormon Tea (*Ephedra trifurca*)
Ocotillo (*Fouquieria splendens*)

**Herbs (Forbs and Grass)**

Cryptantha (*Cryptantha barbigera*)
Fescue (*Festuca*)
Gilia (*Gilia tenuiflora*)
Heron-bill (*Erodium cicutarium*)
Hilaria (*Hilaria rigida*)
Lupine (*Lupinus sparsiflorus*)

**Cacti**

Buckhorn Cholla (*Opuntia acanthocarpa*)
Compass Barrel (*Ferocactus acanthodes*)
Fishhook Barrel (*Ferocactus Wislizeni*)
Fishhook Pincushion (*Mammillaria microcarpa*)
Hedgehog Cactus (*Echinocereus Engelmanii*)
Chain Fruit Cholla (*Opuntia fulgida*)
Pencil Cholla (*Opuntia arbuscula*)
Prickley Pear (*Opuntia phaeacantha var. discata*)
Saguaro (*Carnegiea gigantea*)
Staghorn Choilla (*Opuntia versicolor*)
Teddybear Choilla (*Opuntia Bigelovii*)

## Figure A-12

Rancho Westerno Site Species List.*

*Basic taxonomy from T. H. Kearney and R. H. Peebles, *Arizona Flora* (Berkeley: University of California Press, 1964).

170

1. *Combined Estimate Scale*
   5 covering more than ¾ of plot,                    75–100%
     numbers of individuals optional
   4 covering ½ to ¾ of plot,                          50–75%
     numbers of individuals optional
   3 covering ¼ to ½ of plot, and                      25–50%
     individuals very numerous
   2 covering at least ¹⁄₁₀ to ¼ of plot,              10–25%
     or individuals very numerous
   1 abundantly present, but with low cover             1–10%
     value, or sparse but with greater cover value
   + sparsely present with very low cover value          to 1%
2. *Sociability*
   5 herds of individuals
   4 small colonies or extended patches or mats
   3 troops, larger patches, or bunches
   2 groups or small patches of individuals
   1 individual shoots, stems, trunks
3. *Microrelief* (small inequalities within stand or quadrat)
   a. Uniform
      1. Flat
      2. Convex
      3. Concave
   b. Interrupted
      1. Mounded
      2. Pitted
      3. Ridge and swale
      4. Truncated (slips or blowouts)

## Figure A-13.

Analytical Characteristics of Plant Communities and Selected Physical Site Parameters.

## TABLE A-9

*Rurban Fringe Exercise*

*Operational Procedure*

Exercise to be done in teams of two; length of time 2 to 3 days.

*Problem*

To determine the land use and economic geography of a one square mile area in Metropolis county undergoing dynamic changes from rural to urban uses.

*Hypotheses*

1. Land use changes from agricultural to urban uses are taking place rapidly because of metropolitan expansion.
2. Remaining agricultural land is restricted to low-slope and irrigatable land; topography and availability of water are primary determinants.
3. Vacant land is being held for speculation and future urban development.
4. Lack of orderly growth is apparent and urban services are not well established.
5. Conflicts exist between farmers and urbanites because land uses and resulting tax base and land values are not compatible.

*Method of Analysis*

Land use mapping and interviewing procedures.

The data collected will provide the basis for a written report emphasizing the major geographic aspects of the study area. Major points might include:

1. The physical setting——topography, soils, drainage.
2. Land use patterns——analysis of land use categories.
3. Agricultural aspects——commodities produced, what, where, why, how, markets.
4. Present settlement pattern——farmstead and field patterns, urban uses, residential, commercial, industrial, and transportational features.
5. Historical development——sequent occupance, recent and present changes.
6. Future dynamics and the area——active processes, future trends.

Pertinent illustrations might include:

| | |
|---|---|
| 1. Maps | Land use |
| | Topography, slope, or local relief |
| | Drainage |
| | Settlement |
| | Farm boundaries |
| | Road patterns |
| | Changes since 1980 |
| 2. Tables | Land use |
| | Topography——slope |
| | Commodity production |
| | Population |
| | Flow charts |
| | Changes since 1980 |

## TABLE A-9. (cont.)

*Rurban Fringe Exercise.*

*Equipment needed*

Vertical air photo (1 inch = 400 feet).
Land use mapping keys.
Classification of the physical characteristics of the land.
Abney level.
Topographic quadrangle (USGS 1:24,000).

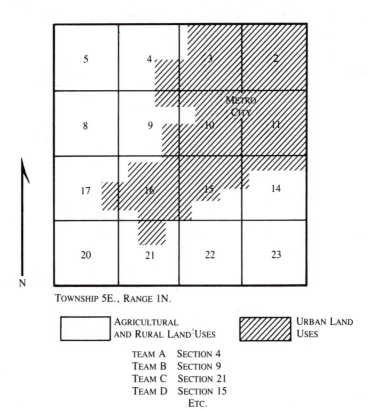

TOWNSHIP 5E., RANGE 1N.

| | AGRICULTURAL AND RURAL LAND USES | | | | URBAN LAND USES |

TEAM A   SECTION 4
TEAM B   SECTION 9
TEAM C   SECTION 21
TEAM D   SECTION 15
ETC.

## Figure A-14.

Locational Map for Hypothetical Rurban Fringe Exercise.

## TABLE A-10.

*Manufacturing Exercise.*

*Operational Procedure*

Exercise to be done individually; length of time 2 days.

*Objectives*

To apply knowledge and skills to the analysis of a specific manufacturing situation. The research design and preparation of the written report are left to the students.

*Problem*

Each student will be assigned a specific industry to analyze. The individual student must establish contacts, arrange for appointments, and design a suitable questionnaire and other appropriate data collecting instruments. A written report is to include the national perspective for this type of industry; historical development; the importance of economic, resource base, social, and governmental locational factors; and relevant site factors. Maps, graphs, tables, flowcharts, etc., should be included.

*Equipment Needed*

Determined by the problem.

## TABLE A-11.

*Individual Exercise.*

*Operational Procedure*

Exercise to be done individually; length of time one week to 10 days.

*Objectives*

To apply knowledge and skills to the solution of an open-ended field research problem.

*Problem*

In the context of the general problem stated below: (1) determine the research design of the overall problem; (2) determine the research matrix, hypotheses, scale, and specific data needed; (3) determine the field data collecting techniques to be employed; (4) complete the field investigations; (5) process and analyze the data; and (6) write a report to include all pertinent maps, tables, etc.

### Problem Statement:

Analyze the recent land use changes in Township X.

### Other Examples of Problem Statements

Compare the historical development and present functions of Village X and Village Y.

Delimit and analyze the CBD of Century City using the Murphy-Vance method, or if not appropriate, devise a workable methodology.

Analyze the commercial land use along Maple Boulevard as to its development and present functions.

Inventory and analyze the vegetation type(s) along the southern slopes of Mt. Highpeak.

Inventory and analyze the settlement pattern(s) in Township Y.

*Equipment Needed*

Determined by the problem.

# Appendix B
## Plane-Table Mapping

The importance of plane-table mapping in geographic field studies is twofold. First, because adequate base maps may not be available in many microlevel research problems, the researcher must construct a map with sufficient orientation points and lines before specific phenomena can be recorded. In this sense, a plane-table map serves as the skeleton or framework within which detailed information may be added. Second, most geographic field research, regardless of the nature or type of problem, is based on the accurate determination of features, objects, and boundaries. A true picture of the spatial distribution of house types, land use, topographical features, lines of transportation, etc., is dependent upon the accuracy of the orientation or control points or lines established on the base map. Precise determination of these points and lines on any base map, whether a plat map, aerial photograph, plane-table map or topographic map, employs the same techniques as used in plane tabling. For these reasons, a detailed description of how to construct a plane-table map, step by step, is given below. It is relevant only to microlevel studies where precision is required.

Two methods of using the plane table will be described:

1. The making of a traverse without elevation information, and
2. The techniques used when elevation information is needed.

In the former case, an open-sight alidade is employed, and in the latter case, a telescopic alidade.

Before setting up for plane tabling, the researcher must select the proper scale for the map and plan the starting point for mapping. This will insure that the study area will fit the paper size and that at no time will the map run off the edge of the board. Paper is then fixed to the surface and the table is *set up*. It is important that the table be at a convenient working height and in a suitable location. It is not always possible to select the optimum location for the table. However, along a road, the shoulder is much safer than the pavement, and setting up the tripod in softer earth is preferable to a hard surface because it is easier to level and secure the setup.

For a discussion of the set-up procedure, a Johnson table type is described. After the researcher decides on the location of the tripod, the legs of the table are set firmly in the ground and the setup aligned over the initial starting point, a surveyed tack point, or reference monument. Next, both wing nuts are loosened; the table is approximately leveled, and the upper nut slightly tightened. Using a circular bubble level, a hand bubble level, or the needle of the compass, the table is leveled as closely as possible, and the upper nut beneath the Johnson table assembly firmly tightened. At this point, the table is level but will rotate in any direction. The greater the degree of leveling of the table, the less time it will take to adjust the telescopic instrument for each sight. If an open-sight instrument is used, it is only necessary to approximately level the table. Next, the table is oriented to magnetic north by laying a compass along one edge and rotating the table until magnetic north is shown. Alternatively, the built-in compass which is found on some tables can be employed for this. The direction of magnetic north should always be marked by a small arrow. After proper orientation, the lower nut on the tripod assembly is locked. It is only necessary to orient the table to magnetic north by using the compass at the initial setup. All subsequent setups are *back-sighted* to bring the table into orientation. A higher degree of accuracy is achieved by using this procedure. The table should not be repeatedly oriented using the compass alone.

To begin mapping, a straight pin is pushed into the table at the desired position of station 1. The edge of the alidade is placed against this pin and the instrument pivoted into the direction on the first sight. The second member of the plane table mapping team then places a sighting pole or rod on the appropriate sight point. The alidade is then carefully aimed at the sight, keeping the straightedge resting against the pin, and a line is drawn along the straightedge toward the next station. This is called a *ray* and represents the line of sight. For proper mapping, both a ray as well as a distance are required to accurately fix the location of the sight station on the map. Therefore, the next step is to obtain the distance, which is accomplished by pacing, chaining, or reading stadia. Pacing is not an accurate method, and therefore is adequate only for some crude surveys. A better method is the use of a tape or a chain, while the easiest involves the use of a telescopic alidade for reading the distance directly.

Once the distance to the second station is ascertained, the engineer's scale is used in conjunction with the map scale previously chosen and a distance marked on the ray that is in porportion to the true distance on the ground. The intersection of this distance and the ray marks the sight location. This point could be called station 2 (see Figure 3-11).

Sightings on any objects within the traverse area or along the traverse line itself may simply be made by sighting with the alidade from the pin position (where the table is located) and drawing a ray. It is not necessary to obtain a distance with the ray if two rays are used from different locations. The intersection of two rays will mark the location of the particular point. This technique is called *intersection,* or *triangulation* (see Figure 3-11). It is a valuable method of establishing points where distance measurements would be difficult. Intersection cannot be done along the traverse line itself because distances there must be measured directly. After rays have been drawn to any interior points of interest, and after the station 2 position has been located, it is possible to remove the table and set up on the next station.

For all secondary setups such as this, the field team needs only to level the table and then reorient it to north using a back-sight technique. To *back-sight,* the straight pin is placed in the station 2 position and the alidade edge along the line stretching from the straight pin back to the original station. The entire table is turned so that the alidade is pointed back to the sighting rod placed on the initial point (station 1), and the table wing nut is tightened. The plane table is now reoriented to north, and the assistant carrying the rod may move ahead and give a sight for another station. This procedure is repeated until the traverse is complete. In most instances, it is best to lay out a *closed traverse* which ends at the same point from which it started. This allows the mapping team to check in at the original point and ascertain any angular or distance errors that might have been made and either correct them or evenly distribute the error back over the stations. The table must never be kicked or leaned on since such stress can change the angle and orientation and result in errors.

The note-taking procedures for elevation mapping using the telescopic alidade are slightly different from the planimetric traverse procedures. When using elevations, there are again initial setup points and new instrument positions; however, instead of moving point to point along the traverse, a leapfrog procedure is the best method (see Figure B-1A). At the initial setup, a forward point called a *turn* point is set up about half-way between the next anticipated instrument position and the present instrument position. This is the elevation control point and is the first point established after the rod is set when the plane table is reoriented and a new rod reading taken from the back-sight. When doing elevation traverses, it is never necessary to take elevation readings at plane-table positions. Therefore, on the notes, the stations are numbered and the setup positions are marked in alphamerics. At each instrument setup, it is necessary to show only the height of the instrument if a Linker rod is not used. This is calculated and is, in fact, the height of the line of sight running through the center of the telescope when it is in a level position. Figure B-1A and B shows an initial elevation of 2500 feet; this is station 1.

Instrument setup *A* is slightly uphill from the initial point of elevation reference. The instrument is leveled and 10 feet is read on the rod. Thus, if the bench mark is 2500 feet, the instrument height is 2500 plus 10 feet, or 2510 feet. The turn position for station 2 is set up just to the right (in Figure B-1B) of instrument station A. The rod reading here is 1.5 feet, which means that the height of the ground at

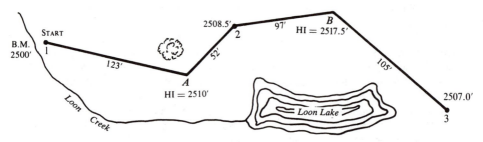

A. Traverse Produced by Using the Telescopic Alidade. Note the "Leapfrog" Station and Instrument Setups.

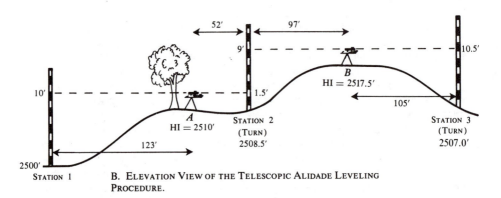

B. Elevation View of the Telescopic Alidade Leveling Procedure.

**Figure B-1.**

Traverse and Elevation Mapping.

station 2 is 1.5 feet subtracted from the height of the instrument (2508.5 feet). This is the turn point, and the elevation is recorded on the map. Also, if possible, the turn is recorded on the ground with a large lumber crayon called a *keel* or in chalk for easy reference. The location of the next instrument set up, point *B*, is determined using normal traverse and plane-table methods without elevations. In other words, a ray and a distance to the next plane-table location are required. The turn position also has a distance and a ray associated with it. Distances are quickly obtained using the stadia interval. After plotting instrument position *B* and making all other necessary sights, the instrument may then be moved to point *B*. After leveling and back-sighting, another level rod reading of 9 feet is made. Since the ground elevation at the turn point (station 2) is 2508.5 feet, the 9-foot reading is added to 2508.5, giving an instrument height of 2517.5 feet. This leapfrog procedure with instrument setups between turn positions is the easiest and most accurate leveling method. At the end of the traverse, it is important to check back into the initial bench mark to determine the extent of any leveling errors.

      In many instances, the ground configuration does not allow sights to be made over very long distances without running off either the bottom or the top of the rod. In these cases, there are several methods that might be employed to continue the plane tabling. One of these is called *stepping*. To step up, the viewing field through the telescope is observed and an object is found which also lies along the upper stadia mark. The telescope is then elevated using the Stebinger screw so that the lower stadia mark is placed on the same reference object. The telescope now has been moved upwards one stadia interval. If the rod is in view and the horizontal cross hair lies on a number, it is possible to read the elevation and add to it the stadia interval. Then the total is added to the height of the instrument. This procedure gives the ground elevation at the rod location. In this example, an addition was performed; however, if the telescopic view is initially above the rod, the sum of the rod reading and the stepped vertical distance must be subtracted from the height of the instrument to obtain ground elevation.

Steppings of more than one stadia interval are possible using other objects which happen to lie within the field of view. Gravel roads provide a multitude of visible objects from which to do stepping.

A similar stepping up or down to get back on the rod may be accomplished by using the Stebinger screw alone. With the numbered scale as a reference, one complete turn will raise or lower the telescope exactly one stadia interval. By repeated rotations, multiple stadia intervals are possible.

Another method for stepping is used when the vertical angles are over a few degrees. This is the Beaman stadia arc method. Here, the vertical angle to the rod reading is measured, and a stadia correction and elevation may be calculated or obtained from published tables.

# Appendix C
## Checklist of Secondary and Tertiary Sources

---

The field researcher must be skilled in the techniques by which primary data are acquired. In addition, the researcher should be familiar with the methods of obtaining information for a given field study from existing published sources. At the worst, secondary and tertiary sources provide the researcher with an overall perspective of the research area. At best, certain published materials may have accuracy, scale, and data such that they can be used directly in the research investigation.

The amount and kinds of secondary and tertiary sources will vary tremendously from one research area to another. The problem is twofold: discovering what has been published and ascertaining where the pertinent documents may be studied. It is beyond the scope of this text to list all possible sources of relevant published information. The sources mentioned are to be considered as a general checklist, and often then sources will refer to or are based on other sources, which may be consulted.

## GOVERNMENT DOCUMENTS

### Federal Government

The United States government is the world's largest publisher. A great number of departments, bureaus, divisions, services, etc., publish materials that are highly significant to geographic field research. Among the several hundreds, the most relevant to conventional geographic field study of an area are:

> *Department of Agriculture*
>> Agricultural Stabilization and Conservation Service
>> Agricultural Research Service
>> Forest Service
>> Soil Conservation Service
>
> *Department of the Army*
>> Corps of Engineers
>
> *Department of Commerce*
>> Bureau of the Census
>> Domestic and International Business Administration
>> National Oceanic and Atmospheric Administration
>> Regional Commission (appropriate commission for specific area)
>
> *Department of the Interior*
>> Fish and Wildlife Service
>> Geological Survey
>> Bureau of Land Management
>> Bureau of Mines
>> National Park Service
>> Bureau of Reclamation
>
> *Department of Transportation*
>> Federal Aviation Administration
>> Federal Highway Administration

### State Government

State departments and agencies publish a great variety of materials pertinent to geographic field study of an area. This information varies greatly in amount and kind from one state to another. However, most states have a department or agency that publishes information in regard to:

> Agriculture and horticulture
> Community development
> Economic planning and development
> Education
> Environment and natural development
> Fish, game, and wildlife
> Highways and transportation
> Historical and archaeological sites
> Humanities and cultural affairs

Indian affairs
Industrial development
Land use and/or land development
Mineral resources and geology
Outdoor education, recreation, and parks
Population distribution and characteristics
Power and utility
Tourism
Water use and management

## Local Government

Often pertinent information to a geographic field research study of a local area is published by the county, cities, and local communities. This information varies in kind and amount, but usually covers the following areas:

Agriculture
Community development and services
Economic and industrial development
Education
Flood control and water use
Highways and transportation
Medical and health services
Parks and recreation
Planning and zoning
Population characteristics
Resources and environment

## PUBLICATIONS OF PROFESSIONAL ORGANIZATIONS

There are over two hundred geographical periodicals or serials published by professional organizations in the United States alone. In addition, approximately one hundred eighty geographical periodicals are published in English by Australian, United Kingdom, and New Zealand professional societies. These publications contain papers concerned with field methodologies and techniques, as well as articles that focus on specific areas. In general, these periodicals may be grouped into the following categories:

Publications of national professional organizations
Publications of regional or divisional professional organizations
Publications of state academies of science or geographical associations
Departments of geography research or discussion papers and monographs

## OTHER PUBLISHED SOURCES

Materials concerning a specific area may be published by local concerns of organizations. Such materials may be of value in that they focus on a small area and often are of a scale and data to supplement the primary data obtained in the field. Such information may be published by:

Banks and real estate firms
Chambers of commerce
Colleges and universities
Corporations and industries
Environmental organizations
Foundations
Historical societies
Industrial development corporations
The League of Women Voters
Newspapers and area magazines
Political groups
Service clubs and organizations
Special commissions
Tourist groups
Utility companies
Water and/or conservancy districts
Wildlife and nature groups

## UNPUBLISHED SOURCES

A wide range of unpublished information usually is available for a specific area. This information may be located in files of government organizations and private corporations. Also, unpublished information may exist in the files of individuals or groups associated with local colleges and universities. The amount and types of such data vary greatly from one area to another.

## BIBLIOGRAPHIC REFERENCES

The bibliography here is representative and not exhaustive. It includes listings of government documents, publications of professional organizations, and bibliographies concerned with geographic field materials.

ANDROIT, J. L., ed., *Guide to U.S. Government Publications.* McLean, Virginia: Documents Index, 1973.
*CURRENT GEOGRAPHICAL PUBLICATIONS.* New York: The American Geographical Society of New York. Monthly except for July and August.
DURRENBERGER, R. W., *Geographical Research and Writing.* New York: Thomas Y. Crowell Co., 1971.
FRIBERG, J. C., *Field Work Techniques: A Revised Research Bibliography for the Fieldworker and Reference Guide for Classroom Studies*, Syracuse University, Department of Geography, Discussion Paper No. 14, 1976.
*GEOGRAPHICAL BIBLIOGRAPHY FOR AMERICAN COLLEGES*, Commission on College Geography, Publication No. 9. Washington, D.C.: Association of American Geographers, 1970.

HARRIS, C. D. annotated *World List of Selected Current Geographical Serials in English, French, and German*, Department of Geography Research Paper No. 137. Chicago: University of Chicago Press, 1971.

HARRIS, C. D., *Bibliography of Geography, Part I: Introduction to General Aids*, Department of Geography Research Paper No. 179. Chicago: University of Chicago Press, 1976.

HARRIS, C. D. and J. D. FELLMANN, *International List of Geographical Series*, Department of Geography Research Paper No. 138. Chicago: University of Chicago, 1971.

McMANIS, D. R., *Historical Geography in the United States: A Bibliography.* Ypsilanti, Michigan: Division of Field Services, Eastern Michigan University, 1965.

*PRICE LISTS.* U.S. Superintendent of Documents. Washington, D.C.: Government Printing Office, Nos. 1-86. List frequently revised. Lists most relevant to field research include:

No. 15 Geology;

No. 25 Transportation, highways, roads, and postal services;

No. 35 National parks;

No. 42 Irrigation, drainage, and water power;

No. 43 Forestry;

No. 46 Soils and fertilizers;

No. 48 Weather, astronomy, and meteorology;

No. 53 Maps;

No. 62 Commerce;

No. 70 Census.

VINGE, C. L. and A. G. VINGE, *U.S. Government Publications for Research and Teaching in Geography and Related Social and Natural Sciences.* Totowa, New Jersey: Littlefield, Adams & Co., 1967.

# Appendix D
## Table of Random Numbers

In geographic field research studies where sampling procedures are employed, a table of random numbers is an essential tool. A table of random numbers may be used to select a sample from any universe which is or can be numbered. The selection of a sample may be made in a number of ways. The key to properly using such a table is systematic selection of numbers. Several examples will serve to demonstrate the use of this table. These examples use Table D-1.

1. The least complex method is to use the same number of digits in a line as the total number of the universe. For example, if the total universe is 900, the digits in any given line (horizontal listing) are divided into groups of three digits each. The starting point may be determined at random, but it is convenient to begin at the start of any line and proceed from left to right. Assuming that a sample of 50 is desired in a universe of 900, and that the starting point is line 25, the sample numbers are 024, 883, 306, 228, 834, 073, and 511. The next group of three digits is 973, which is larger than the total universe. This group is ignored and the next group, 192, is selected. The process continues until 50 sample numbers are defined.

2. If the research design calls for a stratified random sample, random $x$ and $y$ coordinates are required for allocating a sample point to each cell. If the cells are 2.5 inches in width and height, these random numbers can vary between

0.0 and 2.5 if the minimum measurement unit is an even tenth of an inch. Thus, a pair of two-digit numbers in the form x.x are required. Number selections may begin anywhere in the table. However, for the purposes of demonstration, let us assume that the first two numbers to the right of each five-number group are used. Starting at the extreme right of the table and working back across line by line, the numbers selected are:

| x | y |
|---|---|
| 0.0 | 0.7 |
| 1.1 | 1.1 |
| 0.5 | 1.5 |
| 0.2 | 0.9 |
| 1.7 | 0.4 |

etc.

Note that 9.1, 7.0, and 6.9, which appear in number groups 13, 12, and 11 of the first row, are too large for use, so they are discarded, as are 9.0, 9.4, 7.9, 4.6, 4.7, 3.6, and 8.0. Thus, in line 2 only number groups 14, 10, 4, and 2 fall within the range required and are used.

This process is continued until enough numbers are collected. If the table is "used up" going one way, it may be reused by selecting another number pair from the groups of five and using another selection scheme.

3. Some random numbers only require that single digits be selected. For example, if field interviews are being conducted at one in every ten households, numbers 1, 2, 3, 4, 5, 6, 7, 8, 9, and 0 will be used. The selection may be made using the first number in each group of five and then proceeding downward in the table. In this scheme, the following random numbers will be used: 1, 2, 2, 4, 3, 7, 9, 9, 8, 8, 2, 6, 0, 1, 0, 5, 0, 0, 5, 0, 4, 5, 3, 2, 0, etc. (the first twenty-five lines of the first number in the first column of five-digit numbers). Any other selection scheme—diagonal, every other column or row, forward or backward, etc.—may be employed as long as it is systematic. The only stipulation is that a regular selection method be employed.

## TABLE D-1.

*Random Numbers.* *

| Line/Col. | (1) | (2) | (3) | (4) | (5) | (6) | (7) | (8) | (9) | (10) | (11) | (12) | (13) | (14) |
|---|---|---|---|---|---|---|---|---|---|---|---|---|---|---|
| 1 | 10480 | 15011 | 01536 | 02011 | 81647 | 91646 | 69179 | 14194 | 62590 | 36207 | 20969 | 99570 | 91291 | 90700 |
| 2 | 22368 | 46573 | 25595 | 85393 | 30995 | 89198 | 27982 | 53402 | 93965 | 34095 | 52666 | 19174 | 39615 | 99505 |
| 3 | 24130 | 48360 | 22527 | 76393 | 76393 | 64809 | 15179 | 24830 | 49340 | 32081 | 30680 | 19655 | 63348 | 58629 |
| 4 | 42167 | 93003 | 06243 | 61680 | 07856 | 16376 | 39440 | 53537 | 71341 | 57004 | 00849 | 74917 | 97758 | 16379 |
| 5 | 37570 | 39975 | 81837 | 16656 | 06121 | 91782 | 60468 | 81305 | 49684 | 60672 | 14110 | 06927 | 01263 | 54613 |
| 6 | 77921 | 06907 | 11008 | 42751 | 27756 | 53498 | 18602 | 70659 | 90655 | 15053 | 21916 | 81825 | 44394 | 42880 |
| 7 | 99562 | 72905 | 56420 | 69994 | 98872 | 31016 | 71194 | 18738 | 44013 | 48840 | 63213 | 21069 | 10634 | 12952 |
| 8 | 96301 | 91977 | 05463 | 07972 | 18876 | 20922 | 94595 | 56869 | 69014 | 60045 | 18425 | 84903 | 42508 | 32307 |
| 9 | 89579 | 14342 | 63661 | 10281 | 17453 | 18103 | 57740 | 84378 | 25331 | 12566 | 58678 | 44947 | 05585 | 56941 |
| 10 | 85475 | 36857 | 53342 | 53988 | 53060 | 59533 | 38867 | 62300 | 08158 | 17983 | 16439 | 11458 | 18593 | 64952 |
| 11 | 28918 | 69578 | 88231 | 33276 | 70997 | 79936 | 56865 | 05859 | 90106 | 31595 | 01547 | 85590 | 91610 | 78188 |
| 12 | 63553 | 40961 | 48235 | 03427 | 49626 | 69445 | 18663 | 72695 | 52180 | 20847 | 12234 | 90511 | 33703 | 90322 |
| 13 | 09429 | 93969 | 52636 | 92737 | 88974 | 33488 | 36320 | 17617 | 30015 | 08272 | 84115 | 27156 | 30613 | 74952 |
| 14 | 10365 | 61129 | 87529 | 85689 | 48237 | 52267 | 67689 | 93394 | 01511 | 26358 | 85104 | 20285 | 29975 | 89868 |
| 15 | 07119 | 97336 | 71048 | 08178 | 77233 | 13916 | 47564 | 81056 | 97735 | 85977 | 29372 | 74461 | 28551 | 90707 |
| 16 | 51085 | 12765 | 51821 | 51259 | 77452 | 16308 | 60756 | 92144 | 49442 | 53900 | 70960 | 63990 | 75601 | 40719 |
| 17 | 02368 | 21382 | 52404 | 60268 | 89368 | 19885 | 55322 | 44819 | 01188 | 65255 | 64835 | 44919 | 05944 | 55157 |
| 18 | 01011 | 54092 | 33362 | 94904 | 31273 | 04146 | 18594 | 29852 | 71585 | 85030 | 51132 | 01915 | 92747 | 64951 |
| 19 | 52162 | 53916 | 46369 | 58586 | 23216 | 14513 | 83149 | 98736 | 23495 | 64350 | 94738 | 17752 | 35156 | 35749 |
| 20 | 07056 | 97628 | 33787 | 09998 | 42698 | 06691 | 76988 | 13602 | 51851 | 46104 | 88916 | 19509 | 25625 | 58104 |
| 21 | 48663 | 91245 | 85828 | 14346 | 09172 | 30168 | 90229 | 04734 | 59193 | 22178 | 30421 | 61666 | 99904 | 32812 |
| 22 | 54164 | 58492 | 22421 | 74103 | 47070 | 25306 | 76468 | 26384 | 58151 | 06646 | 21524 | 15227 | 96909 | 44592 |
| 23 | 32639 | 32363 | 05597 | 24200 | 13363 | 38005 | 94342 | 28728 | 35806 | 06912 | 17012 | 64161 | 18296 | 22851 |
| 24 | 29334 | 27001 | 87637 | 87308 | 58731 | 00256 | 45834 | 15398 | 46557 | 41135 | 10367 | 07684 | 36188 | 18510 |
| 25 | 02488 | 33062 | 28834 | 07351 | 19731 | 92420 | 60952 | 61280 | 50001 | 67658 | 32586 | 86679 | 50720 | 94953 |

## TABLE D-1. (cont.)

*Random Numbers.**

| Line/Col. | (1) | (2) | (3) | (4) | (5) | (6) | (7) | (8) | (9) | (10) | (11) | (12) | (13) | (14) |
|---|---|---|---|---|---|---|---|---|---|---|---|---|---|---|
| 26 | 81525 | 72295 | 04839 | 96423 | 24878 | 82651 | 66566 | 14778 | 76797 | 14780 | 13300 | 87074 | 79666 | 95725 |
| 27 | 29676 | 20591 | 68086 | 26432 | 46901 | 20849 | 89768 | 81536 | 86645 | 12659 | 92259 | 57102 | 80428 | 25280 |
| 28 | 00742 | 57392 | 39064 | 66432 | 84673 | 40027 | 32832 | 61362 | 98947 | 96067 | 64760 | 64584 | 96096 | 98253 |
| 29 | 05366 | 04213 | 25669 | 26422 | 44407 | 44048 | 37937 | 63904 | 45766 | 66134 | 75470 | 66520 | 34693 | 90449 |
| 30 | 91921 | 26418 | 64117 | 94305 | 26766 | 25940 | 39972 | 22209 | 71500 | 64568 | 91402 | 42416 | 07844 | 69618 |
| 31 | 00582 | 04711 | 87917 | 77341 | 42206 | 35126 | 74087 | 99547 | 81817 | 42607 | 43808 | 76655 | 62028 | 76630 |
| 32 | 00725 | 69884 | 62797 | 56170 | 86324 | 88072 | 76222 | 36086 | 84637 | 93161 | 76038 | 65855 | 77919 | 88006 |
| 33 | 69011 | 65795 | 95876 | 55293 | 18988 | 27354 | 26575 | 08625 | 40801 | 59920 | 29841 | 80150 | 12777 | 48501 |
| 34 | 25976 | 57948 | 29888 | 88604 | 67917 | 48708 | 18912 | 82271 | 65424 | 69774 | 33611 | 54262 | 85963 | 03547 |
| 35 | 09763 | 83473 | 73577 | 12908 | 30883 | 18317 | 28290 | 35797 | 05998 | 41688 | 34952 | 37888 | 38917 | 88050 |
| 36 | 91567 | 42595 | 27958 | 30134 | 04024 | 86385 | 29880 | 99730 | 55536 | 84855 | 29080 | 09250 | 79656 | 73211 |
| 37 | 17955 | 56349 | 90999 | 49127 | 20044 | 59931 | 06115 | 20542 | 18059 | 02008 | 73708 | 83517 | 36103 | 42791 |
| 38 | 46503 | 18584 | 18845 | 49618 | 02304 | 51038 | 20655 | 58727 | 28168 | 15475 | 56942 | 53389 | 20562 | 87338 |
| 39 | 92157 | 89634 | 94824 | 78171 | 84610 | 82834 | 09922 | 25417 | 44137 | 48413 | 25555 | 21246 | 35509 | 20468 |
| 40 | 14577 | 62765 | 35605 | 81263 | 39667 | 47358 | 56873 | 56307 | 61607 | 49518 | 89656 | 20103 | 77490 | 18062 |
| 41 | 98427 | 07523 | 33362 | 64270 | 01638 | 92477 | 66969 | 98420 | 04880 | 45585 | 46565 | 04102 | 46880 | 45709 |
| 42 | 34914 | 63976 | 88720 | 82765 | 34476 | 17032 | 87589 | 40836 | 32427 | 70002 | 70663 | 88863 | 77775 | 69348 |
| 43 | 70060 | 28277 | 39475 | 46473 | 23219 | 53416 | 94970 | 25832 | 69975 | 94884 | 19661 | 72828 | 00102 | 66794 |
| 44 | 53976 | 54914 | 06990 | 67245 | 68350 | 82948 | 11398 | 42878 | 80287 | 88267 | 47363 | 46634 | 06541 | 97809 |
| 45 | 76072 | 29515 | 40980 | 07391 | 58745 | 25774 | 22987 | 80059 | 39911 | 96189 | 41151 | 14222 | 60697 | 59583 |
| 46 | 90725 | 52210 | 83974 | 29992 | 65831 | 38857 | 50490 | 83765 | 55657 | 14361 | 31720 | 57375 | 56228 | 41546 |
| 47 | 64364 | 67412 | 33339 | 31926 | 14883 | 24413 | 59744 | 92351 | 97473 | 89286 | 35931 | 04110 | 23726 | 51900 |
| 48 | 08962 | 00358 | 31662 | 25388 | 61642 | 34072 | 81249 | 35648 | 56891 | 69352 | 48373 | 45578 | 78547 | 81788 |
| 49 | 95012 | 68379 | 93526 | 70765 | 10592 | 04542 | 76463 | 54328 | 02349 | 17247 | 28865 | 14777 | 62730 | 92277 |
| 50 | 15664 | 10493 | 20492 | 38391 | 91132 | 21999 | 59516 | 81652 | 27195 | 48223 | 46751 | 22923 | 32261 | 85653 |
| 51 | 16408 | 81899 | 04153 | 53381 | 79401 | 21438 | 83035 | 92350 | 36693 | 31238 | 59649 | 91754 | 72772 | 02338 |
| 52 | 18629 | 81953 | 05520 | 91962 | 04739 | 13092 | 97662 | 24822 | 94730 | 06496 | 35090 | 04822 | 86774 | 98289 |
| 53 | 73115 | 35101 | 47498 | 87637 | 99016 | 71060 | 88824 | 71013 | 18735 | 20286 | 23153 | 72924 | 35165 | 43040 |
| 54 | 57491 | 16703 | 23167 | 49323 | 45021 | 33132 | 12544 | 41035 | 80780 | 45393 | 44812 | 12515 | 98931 | 91202 |
| 55 | 30405 | 83946 | 23792 | 14422 | 15059 | 45799 | 22716 | 19792 | 09983 | 74353 | 68668 | 30429 | 70735 | 25499 |
| 56 | 16631 | 35006 | 85900 | 98275 | 32388 | 52390 | 16815 | 69298 | 82732 | 38480 | 73817 | 32523 | 41961 | 44437 |
| 57 | 96773 | 20206 | 42559 | 78985 | 05300 | 22164 | 24369 | 54224 | 35083 | 19687 | 11052 | 91491 | 60383 | 19746 |
| 58 | 38935 | 64202 | 14349 | 82674 | 66523 | 44133 | 00697 | 35552 | 35970 | 19124 | 63318 | 29686 | 03387 | 59846 |
| 59 | 31624 | 76384 | 17403 | 53363 | 44167 | 64486 | 64758 | 75366 | 76554 | 31601 | 12614 | 33072 | 60332 | 92325 |
| 60 | 78919 | 19474 | 23632 | 27889 | 47914 | 02584 | 37680 | 20801 | 72152 | 39339 | 34806 | 08930 | 85001 | 87820 |
| 61 | 03931 | 33309 | 57047 | 74211 | 63445 | 17361 | 62825 | 39908 | 05607 | 91284 | 68833 | 25570 | 38818 | 46920 |
| 62 | 74426 | 33278 | 43972 | 10119 | 89917 | 15665 | 52872 | 73823 | 73144 | 88662 | 88970 | 74492 | 51805 | 99378 |
| 63 | 09066 | 00903 | 20795 | 95452 | 92648 | 45454 | 09552 | 88815 | 16553 | 51125 | 79375 | 97596 | 16296 | 66092 |
| 64 | 42238 | 12426 | 87025 | 14267 | 20979 | 04508 | 64535 | 31355 | 86064 | 29472 | 47689 | 05974 | 52468 | 16834 |
| 65 | 16153 | 08002 | 26504 | 41744 | 81959 | 65642 | 74240 | 56302 | 00033 | 67107 | 77510 | 70625 | 28725 | 34191 |
| 66 | 21457 | 40742 | 29820 | 96783 | 29400 | 21840 | 15035- | 34537 | 33310 | 06116 | 95240 | 15957 | 16572 | 06004 |
| 67 | 21581 | 57802 | 02050 | 89728 | 17937 | 37621 | 47075 | 42080 | 97403 | 48626 | 68995 | 43805 | 33386 | 21597 |
| 68 | 55612 | 78095 | 83197 | 33732 | 05810 | 24813 | 86902 | 60397 | 16489 | 03264 | 88525 | 42786 | 05269 | 92532 |
| 69 | 44657 | 66999 | 99324 | 51281 | 84463 | 60563 | 79312 | 93454 | 68876 | 25471 | 93911 | 25650 | 12682 | 73572 |
| 70 | 91340 | 84979 | 46949 | 81973 | 37949 | 61023 | 43997 | 15263 | 80644 | 43942 | 89203 | 71795 | 99533 | 50501 |
| 71 | 91227 | 21199 | 31935 | 27022 | 84067 | 05462 | 35216 | 14486 | 29891 | 68607 | 41867 | 14951 | 91696 | 85065 |
| 72 | 50001 | 38140 | 66321 | 19924 | 72163 | 09538 | 12151 | 06878 | 91903 | 18749 | 34405 | 56087 | 82790 | 70925 |
| 73 | 65390 | 05224 | 72958 | 28609 | 81406 | 39147 | 25549 | 48542 | 42627 | 45233 | 57202 | 94617 | 23772 | 07896 |
| 74 | 27504 | 96131 | 83944 | 41575 | 10573 | 08619 | 64482 | 73923 | 36152 | 05184 | 94142 | 25299 | 84387 | 34925 |
| 75 | 37169 | 94851 | 39117 | 89632 | 00959 | 16487 | 65536 | 49071 | 39782 | 17095 | 02330 | 74301 | 00275 | 48280 |
| 76 | 11508 | 70225 | 51111 | 38351 | 19444 | 66499 | 71945 | 05422 | 13442 | 78675 | 84081 | 66938 | 93654 | 59894 |
| 77 | 37449 | 30362 | 06694 | 54690 | 04052 | 53115 | 62757 | 95348 | 78662 | 11163 | 81651 | 50245 | 34971 | 52924 |
| 78 | 46515 | 70331 | 85922 | 38329 | 57015 | 15765 | 97161 | 17869 | 45349 | 61796 | 66345 | 81073 | 49106 | 79860 |
| 79 | 30986 | 81223 | 42416 | 58353 | 21532 | 30502 | 32305 | 86482 | 05174 | 07901 | 54339 | 58861 | 74818 | 46942 |
| 80 | 63798 | 64995 | 46583 | 09785 | 44160 | 78128 | 83991 | 42865 | 92520 | 83531 | 80377 | 35909 | 81250 | 54238 |
| 81 | 82486 | 84846 | 99254 | 67632 | 43218 | 50076 | 21361 | 64816 | 51202 | 88124 | 41870 | 52689 | 51275 | 83556 |
| 82 | 21885 | 32906 | 92431 | 09060 | 64297 | 51674 | 64126 | 62570 | 26123 | 05155 | 59194 | 52799 | 28225 | 85762 |
| 83 | 60336 | 98782 | 07408 | 53458 | 13564 | 59089 | 26445 | 29789 | 85205 | 41001 | 12535 | 12133 | 14645 | 23541 |
| 84 | 43937 | 46891 | 24010 | 25560 | 86355 | 33941 | 25786 | 54990 | 71899 | 15475 | 95434 | 98227 | 21824 | 19585 |
| 85 | 97656 | 63175 | 89303 | 16275 | 07100 | 92063 | 21942 | 18611 | 47348 | 20203 | 18534 | 03862 | 78095 | 50136 |
| 86 | 03299 | 01221 | 05418 | 38982 | 55758 | 92237 | 26759 | 86367 | 21216 | 98442 | 08303 | 56613 | 91511 | 75928 |
| 87 | 79626 | 06486 | 03574 | 17668 | 07785 | 76020 | 79924 | 25651 | 83325 | 88428 | 85076 | 72811 | 22717 | 50585 |
| 88 | 85636 | 68335 | 47539 | 03129 | 65651 | 11977 | 02510 | 26113 | 99447 | 68645 | 34327 | 15152 | 55230 | 93448 |
| 89 | 18039 | 14367 | 61337 | 06177 | 12143 | 46609 | 32989 | 74014 | 64708 | 00533 | 35398 | 58408 | 13261 | 47903 |
| 90 | 08362 | 15656 | 60627 | 36478 | 65648 | 16764 | 53412 | 09013 | 07832 | 41574 | 17639 | 82163 | 60859 | 75567 |
| 91 | 79556 | 29068 | 04142 | 16268 | 15387 | 12856 | 66227 | 38358 | 22478 | 73373 | 88732 | 09443 | 82558 | 05250 |
| 92 | 92608 | 82674 | 27072 | 32534 | 17075 | 27698 | 98204 | 63863 | 11951 | 34648 | 88022 | 56148 | 34925 | 57031 |
| 93 | 23982 | 25835 | 40055 | 67006 | 12293 | 02753 | 14827 | 23235 | 35071 | 99704 | 37543 | 11601 | 35503 | 85171 |
| 94 | 09915 | 96306 | 05908 | 97901 | 28395 | 14186 | 00821 | 80703 | 70426 | 75647 | 76310 | 88717 | 37890 | 40129 |
| 95 | 59037 | 33300 | 26695 | 62247 | 69927 | 76123 | 50842 | 43834 | 86654 | 70959 | 79725 | 93872 | 28117 | 19233 |
| 96 | 42488 | 78077 | 69882 | 61657 | 34136 | 79180 | 97526 | 43092 | 04098 | 73571 | 80799 | 76536 | 71255 | 64239 |
| 97 | 46764 | 86273 | 63003 | 93017 | 31204 | 36692 | 40202 | 35275 | 57306 | 55543 | 53203 | 18098 | 47625 | 88684 |
| 98 | 03237 | 45430 | 55417 | 63282 | 90816 | 17349 | 88298 | 90183 | 36000 | 78406 | 06216 | 95787 | 42579 | 90730 |
| 99 | 86591 | 81482 | 52667 | 61582 | 14972 | 90053 | 89534 | 76036 | 49199 | 43716 | 97548 | 04379 | 46370 | 28672 |
| 100 | 38534 | 01715 | 94964 | 87288 | 65680 | 43772 | 39560 | 12918 | 86537 | 62738 | 19636 | 51132 | 25739 | 56947 |

*Reprinted with permission from *Handbook of Tables for Probability and Statistics*, 2nd edition, edited by William H. Beyer (Cleveland: The Chemical Rubber Co., 1968). Copyright, The Chemical Rubber Co., CRC Press, Inc.

# Appendix E
## Land Use Classification Systems

Many field research problems require classifying land uses in some manner. At this time no standardized classification system like the ones for vegetation, animals, minerals, or rocks yet exists. This lack of uniform system is due, in part, to the great variety of land use studies concerning scale, research areas, and objectives. Nonetheless several systems (similar in nature) are used frequently; the researcher may either use an existing system as is, modify an existing classification to fit a given field study, or design a new system tailored to a particular study or area. In selecting or designing a classification system, the researcher needs to consider

1. the phenomena to be mapped or recorded,
2. the scale of the field work to be accomplished (size of the minimum area to be mapped),
3. whether single or multi-feature mapping techniques will be employed,
4. how the acquired data will be compiled and processed,
5. the manner in which rotational land uses will be recorded,
6. how to deal with mixed land use within the minimum size mapping area,
7. if nonvisible land use data is to be obtained, what type of interviewing procedures must be established, and
8. who is to do the field work (one person or teams of researchers? What will be the expertise and experience of the proposed field personnel?).

The classification system described has been used successfully (in modified forms) in many sections of the continental United States, Puerto Rico, and other parts of the world. Only the framework of the classification is illustrated here. It is a stratified classification, and data may be collected in varying degrees of detail compatible to the scale of mapping. The researcher may add, or otherwise modify within the existing framework or format, to suit the needs of the study or environment without difficulty.

*First Symbol—Major Land Use*

R. Rural Land    U. Urban Land

## R. Rural Land Use

### Second Symbol—General Land Use

| 1. Cropped Land | 2. Pasture Land | 3. Forest, Grassland, Shrub | 4. Idle Land | 5. Miscellaneous |
|---|---|---|---|---|

### Third Symbol—Specific Land Use[1].

| 1. Cropped Land | 2. Pasture Land[2] | 3. Forest, Grassland, Shrub[3]. | 4. Idle Land | 5. Miscellaneous |
|---|---|---|---|---|
| a. alfalfa | a. rotation pasture | a. spruce-fir | a. waste land | a. airport |
| b. barley | b. permanent, non-wooded | b. yellow pine-douglas fir | b. abandoned farm land | b. school |
| c. cotton | c. woodland pasture | c. pinon pine-juniper | c. held for non-agricultural development | c. cemetery |
| d. dates | d. brushland pasture | d. sagebrush (northern desert shrub) | etc. | d. park |
| e. corn | | e. cresote (southern desert shrub) | | e. mine or quarry |
| f. fallow | | f. greasewood (salt desert shrub) | | f. industry |
| etc. | | g. shortgrass (plains grassland) | | g. golf course |
| | | h. mesquite grass (desert grassland) | | etc. |

### Fourth Symbol—Quality

| 1. Cropped Land[4] | 2. Pasture Land[4] | 3. Forest, Grassland, Shrub | 4. Idle Land | 5. Miscellaneous |
|---|---|---|---|---|
| a. good quality | a. good quality | a. merchantable | (no fourth symbol) | (no fourth symbol) |
| b. moderate | b. moderate | b. potentially merchantable | | |
| c. poor quality | c. poor quality | c. scrub—not merchantable | | |

1. The *Rural land use* mapping key is designed for mapping large areas, e.g., fields, woodlands. For small areas such as farmsteads, buildings, and roads use the RURAL CULTURAL FEATURES KEY.
2. Pasture land may be defined as follows: *rotation pasture*—presently being used as pasture but to be used as a cropped land later; *permanent pasture*—used as pasture year after year; does not fall into the cropped land rotation system; *woodland pasture*—at least 50 percent is covered by trees twelve feet or higher; *brushland pasture*—50 percent or more covered by bush, small trees, etc., under twelve feet in height.
3. Applies only to Southwestern United States and Northern Mexico. This category must be revised to fit the particular area surveyed. Similar subcategories can be developed, however, for all sections of the country and abroad.
4. Quality may be determined subjectively by judging the general condition of the crop or pasture land. If the mapping unit is free of weeds, and bare areas, and the crop is uniform in color and height it may be judged good quality. Quality may also be determined on a measurement basis—e.g., corn yields of 100 bu, per acre—good; 70–100 bu. per acre—moderate; under 70—poor; carrying capacity of pastures, etc.

## Rural Cultural Features Key

### First Symbol—General Functional Use

1. Farmsteads
2. Non-Farm Country Houses
3. Resort Dwellings
4. Communal Buildings
5. Commerical Buildings
6. Manufactural Structures
7. Roads
8. Irrigation or Drainage Ditches
9. Agricultural Structures

### Second Symbol—Specific Functional Use

1. Farmsteads
   a. beef cattle
   b. citrus
   c. dairy
   d. poultry
   e. cash grain
   f. truck

   etc.

2. Non-Farm Houses
   a. large (7 or more rooms)
   b. medium (4 to 6 rooms)
   c. small (less than 4 rooms)

3. Resort Dwellings
   a. year-round
   b. seasonal

4. Communal Buildings
   a. school
   b. church
   c. club house

   etc.

5. Commercial Buildings
   a. general store
   b. food store
   c. clothing store
   d. hardware store
   e. service station
   f. tavern
   g. drive-in theater

   etc.

6. Manufactural Structure
   a. gravel pit buildings and plant
   b. quarry and buildings
   c. cheese factory
   d. butter factory
   e. cannery

   etc.

7. Roads
   a. paved (two lane)
   b. gravel
   c. improved dirt
   d. unimproved
   e. abandoned
   f. farm lane
   g. paved (four lane)

8. Irrigation or Drainage Ditches
   a. less than 5 ft wide
   b. 5 ft.–10 ft. wide
   c. 10 ft.–20 ft. wide
   d. 20 ft. or more

9. Agricultural Structures
   a. loading pen
   b. cotton gin
   c. grainery
   d. cattle vat
   e. creamery

   etc.

### Third Symbol—Age

1. since 1980
2. 1950–1980
3. 1920–1950
4. prior to 1920

(or other criteria suitable to the specific area)

## U. Urban Land Use

### Second Symbol—General Land Use

| 1. Residential | 2. Commercial | 3. Industrial | 4. Governmental and Public Utilities | 5. Institutional | 6. Vacant |
|---|---|---|---|---|---|

### Third Symbol—Specific Land Use

| 1. Residential | 2. Commercial | 3. Industrial | 4. Governmental and Public Utilities | 5. Institutional |
|---|---|---|---|---|
| a. single family | a. appliance store | a. food and kindred products | a. court house | a. school |
| b. two family | b. bank, savings and loan | b. tobacco manufacturers | b. fire station | b. church |
| c. multi-family or apartments | c. cafeteria, restaurant | c. textile mill products | c. police station | c. hospital |
| | d. department store | d. apparel and clothing | d. powerhouse or substation | etc. |
| | e. drug store | e. lumber and wood products (except furniture) | e. parks or recreation | |
| | f. furniture store | f. furniture and fixtures | f. water works | |
| | g. grocery | g. paper and allied products | g. post office | |
| | h. hardware | h. printing, publishing and allied industries | etc. | |
| | i. tavern, bar | i. chemicals and allied products | | |
| | j. jewelry, gifts | j. products of petroleum and coal | | |
| | k. beauty salon | etc. | | |
| | l. men's clothing | | | |
| | m. bakery | | | |
| | etc. | | | |

### Fourth Symbol—General Condition

1. well kept up    2. moderate condition    3. deteriorating
(objective evaluation procedure may be defined compatible to area mapped)

### Fifth Symbol-Age

1. since 1960    2. 1950–1980    3. 1920–1950    4. prior to 1920
(or other criteria suitable to the specific area)

The researcher should also be familiar with two standardized land use classification systems that have been designed to be applicable in a variety of environments—*The Standard Land Use Coding Manual*[1] and *A Land Use and Land Cover Classification System for Use with Remote Sensor Data*.[2] The first employs four levels of categories, and the second uses three levels of categories. Both have nine generalized first level sub-categories. These classifications should be considered before other systems are developed. The basic elements of each of these classifications are illustrated below.

## First Level Categories[1]

1. residential
2. manufacturing (9 second level categories included)
3. manufacturing (6 second level categories included)
4. transportation, communications, and utilities
5. trade
6. services
7. cultural, entertainment, and recreation
8. resource production and extraction
9. undeveloped land and water areas

| Level I | Level II | Level III | Level IV |
|---|---|---|---|
| 6. services | 65. professional services | 651. medical and other health services | 6511. physicians' services |
| | | | 6512. dental services |
| | | | 6513. hospital services |
| | | | 6514. medical laboratory services |
| | | | 6515. dental laboratory services |
| | | | 6516. sanitariums, convalescent, and rest home services |
| | | | 6517. medical clinics—out-patient services |
| | | | 6519. other medical and health services |

First Level Categories[2]

1. urban or built-up land
2. agricultural land
3. rangeland
4. forest land
5. water
6. wetland
7. barren land
8. tundra
9. perennial snow or ice

| Level I | Level II | Level III |
|---------|----------|-----------|
| 1. urban or built up | 11. residential | 111. single-family units |
| | | 112. multi-family units |
| | | 113. group quarters |
| | | 114. residential hotels |
| | | 115. mobile home parks |
| | | 116. transient lodgings |
| | | 117. other |

1. Federal Highway Administration, Bureau of Public Roads. *Standard Land Use Coding Manual.* (Washington: U.S. Government Printing Office, 1969).
2. J. R. Anderson, E.E. Hardy, J.T. Roach, and R.E. Witmer, *A Land Use and Land Cover Classification System for Use with Remote Sensor Data,* Geological Survey Professional Paper 964 (Washington: U.S. Government Printing Office, 1976).

# Appendix F
## Measures and Conversion Tables

---

### 1. Linear Measure

| | | |
|---:|---:|:---|
| 1 mil = 0.001 inch | = | 0.0254 millimeter |
| 1 inch = 1,000 mils | = | 2.54 centimeters |
| 12 inches = 1 foot | = | 0.3048 meter |
| 3 feet = 1 yard | = | 0.9144 meter |
| 5½ yards or 16½ feet = 1 rod (or pole or perch) | = | 5.029 meters |
| 40 rods = 1 furlong | = | 201.168 meters |
| 8 furlongs or 1,760 yards or 5,280 feet = 1 (statute) mile | = | 1.6093 kilometers |
| 3 miles = 1 (land) league | = | 4.83 kilometers |

| | | | |
|---:|:---|---:|:---|
| | 1 millimeter | = 0.03937 | inch |
| 10 millimeters | = 1 centimeter | = 0.3937 | inch |
| 10 centimeters | = 1 decimeter | = 3.937 | inches |
| 10 decimeters | = 1 meter | = 39.37 | inches or 3.2808 feet |
| 10 meters | = 1 decameter | = 393.7 | inches |
| 10 decameters | = 1 hectometer | = 328.08 | feet |
| 10 hectometers | = 1 kilometer | = 0.621 | mile or 3,280.8 feet |
| 10 kilometers | = 1 myriameter | = 6.21 | miles |

---

### 2. Areal (Square) Measure

| | | |
|---:|---:|:---|
| 1 square inch | = | 6.452 square centimeters |
| 144 square inches = 1 square foot | = | 929.03 square centimeters |
| 9 square feet = 1 square yard | = | 0.8361 square meter |
| 30¼ square yards = 1 square rod (or square pole or square perch) | = | 25.292 square meters |
| 160 square rods or 4,840 square yards or 43,560 square feet = 1 acre | = | 0.4047 hectare |
| 640 acres = 1 square mile | = | 259.00 hectares or 2.590 square kilometers |

| | | | |
|---:|:---|---:|:---|
| 1 square millimeter | = 0.00155 | square inch |
| 100 square millimeters = 1 square centimeter | = 0.15499 | square inch |
| 100 square centimeters = 1 square decimeter | = 15.499 | square inches |
| 100 square decimeters = 1 square meter | = 1,549.9 | square inches or 1.196 square yards |
| 100 square meters = 1 square decameter | = 119.6 | square yards |
| 100 square decameters = 1 square hectometer | = 2.471 | acres |
| 100 square hectometers = 1 square kilometer | = 0.386 | square mile or 247.1 acres |

| | | | |
|---:|:---|---:|:---|
| 1 square meter | = 1 centiare | = 1,549.9 | square inches |
| 100 centiares | = 1 are | = 119.6 | square yards |
| 100 ares | = 1 hectare | = 2.471 | acres |
| 100 hectares | = 1 square kilometer | = 0.386 | square mile or 247.1 acres |

---

## 3. Chain Measure

(for Gunter's, or surveyor's, chain)

| | | | |
|---|---|---|---|
| 7.92 inches = 1 link | = | 20.12 | centimeters |
| 100 links or 66 feet = 1 chain | = | 20.12 | meters |
| 10 chains or 220 yards = 1 furlong | = | 201.17 | meters |
| 80 chains = 1 mile | = | 1.6093 | kilometers |

(for engineer's chain)

| | | | |
|---|---|---|---|
| 1 foot = 1 link | = | 0.3048 | meter |
| 100 feet = 1 chain | = | 30.48 | meters |
| 52.8 chains = 1 mile | = | 1,609.3 | meters |

## 4. Surveyor Measure

| | | | | |
|---|---|---|---|---|
| 625 square links | = 1 square pole | = | 25.29 | square meters |
| 16 square poles | = 1 square chain | = | 404.7 | square meters |
| 10 square chains | = 1 acre | = | 0.4047 | hectare |
| 640 acres | = 1 square mile or 1 section | = | 259.00 | hectares or 2.59 square kilometers |
| 36 square miles | = 1 township | = | 9,324.0 | hectares or 93.24 square kilometers |

## 5. Nautical Measure

| | |
|---|---|
| 6 feet = 1 fathom | = 1.829 meters |
| 100 fathoms = 1 cable's length (ordinary) | |
| (In the U.S. Navy 120 fathoms or 720 feet, or 219.456 meters, = 1 cable's length; in the British Navy, 608 feet, or 185.319 meters, = 1 cable's length.) | |
| 10 cables' length = 1 international nautical mile (6,076.11549 feet, by international agreement) | = 1.852 kilometers (exactly) |
| 1 international nautical mile = 1.150779 statute miles (the length of a minute of longitude at the equator) | |
| 3 nautical miles = 1 marine league (3.45 statute miles) | = 5.56 kilometer |
| 60 nautical miles = 1 degree of a great circle of the earth = 69.047 statue miles | |

## 6. Angular/Circular Measure

60 seconds (") = 1 minute (')
60 minutes = 1 degree (°)
90 degrees = 1 quadrant or 1 right angle
180 degrees = 2 quadrants or 1 straight angle
4 quadrants or 360 degrees = 1 circle

## 7. Mass

1 ounce (av) = 28.35 grams
1 gram = 0.0353 ounce (av)
1 kilogram = 2.2046 pounds (av) = 1000 grams
1 metric ton = 1000 kilograms = 2204.6 pounds (av) = 1.10 short tons

## 8. Volume

1 cubic foot = 0.028 cubic meter
1 cubic meter = 1.308 cubic yards = 35.31 cubic feet = 61,024 cubic inches
1 cubic kilometer = 0.2399 cubic mile

## 9. Velocity

1 knot (nautical mile) = 1.1516 statute miles per hour = 0.5144 meter per second = 1.85 kilometers per hour
1 meter per second = 3.281 feet per second = 1.942 nautical miles per hour = 2.237 statute miles per hour = 3.6 kilometers per hour

## 10. Energy/Power

1 calorie = 4.186 joules = $3.968 \times 10^{-3}$ British thermal units
1 joule = 0.738 foot pound = $10^7$ ergs
1 British thermal unit = 251.98 calories = 1055 joules = 0.293 watt-hour
1 langley = 1 calorie per square centimeter
1 horse power = 746 watts = 33,000 foot pounds per minute
1 calorie per minute = 251.16 watts
Solar constant = approximately 2 langleys per minute

## 11. Temperature

Temperature in Celsius = 5/9 (temperature in F –32)
Temperature in Fahrenheit = 9/5 (temperature in C) + 32
Temperature in Kelvin or Absolute = temperature in C –273.15

### Fahrenheit to Celsius (Centigrade)
**The vertical column on the left indicates 10 F intervals; 1 F intervals shown across the top.**

| Degrees Fahrenheit | Degrees Centigrade | | | | | | | | | |
|---|---|---|---|---|---|---|---|---|---|---|
| | 0 | 1 | 2 | 3 | 4 | 5 | 6 | 7 | 8 | 9 |
| 110 | 43.3 | 43.9 | 44.4 | 45.0 | 45.6 | 46.1 | 46.7 | 47.2 | 47.8 | 48.3 |
| 100 | 37.8 | 38.3 | 38.9 | 39.4 | 40.0 | 40.6 | 41.1 | 41.7 | 42.2 | 42.8 |
| 90 | 32.2 | 32.8 | 33.3 | 33.9 | 34.4 | 35.0 | 35.6 | 36.1 | 36.7 | 37.2 |
| 80 | 26.7 | 27.2 | 27.8 | 28.3 | 28.9 | 29.4 | 30.0 | 30.6 | 31.1 | 31.7 |
| 70 | 21.1 | 21.7 | 22.2 | 22.8 | 23.3 | 23.9 | 24.4 | 25.0 | 25.6 | 26.1 |
| 60 | 15.6 | 16.1 | 16.7 | 17.2 | 17.8 | 18.3 | 18.9 | 19.4 | 20.0 | 20.6 |
| 50 | 10.0 | 10.6 | 11.1 | 11.7 | 12.2 | 12.8 | 13.3 | 13.9 | 14.4 | 15.0 |
| 40 | 4.4 | 5.0 | 5.6 | 6.1 | 6.7 | 7.2 | 7.8 | 8.3 | 8.9 | 9.4 |
| 30 | –1.1 | –0.6 | 0.0 | 0.6 | 1.1 | 1.7 | 2.2 | 2.8 | 3.3 | 3.9 |
| 20 | –6.7 | –6.1 | –5.6 | –5.0 | –4.4 | –3.9 | –3.3 | –2.8 | –2.2 | –1.7 |
| 10 | –12.2 | –11.7 | –11.1 | –10.6 | –10.0 | –9.4 | –8.9 | –8.3 | –7.8 | –7.2 |
| + 0 | –17.8 | –17.2 | –16.7 | –16.1 | –15.6 | –15.0 | –14.4 | –13.9 | –13.3 | –12.8 |
| – 0 | –17.8 | –18.3 | –18.9 | –19.4 | –20.0 | –20.6 | –21.1 | –21.7 | –22.2 | –22.8 |
| –10 | –23.3 | –23.9 | –24.4 | –25.0 | –25.6 | –26.1 | –26.7 | –27.2 | –27.8 | –28.3 |
| –20 | –28.9 | –29.4 | –30.0 | –30.6 | –31.1 | –31.7 | –32.2 | –32.8 | –33.3 | –33.9 |
| –30 | –34.4 | –35.0 | –35.6 | –36.1 | –36.7 | –37.2 | –37.8 | –33.3 | –38.9 | –39.1 |
| –40 | –40.0 | –40.6 | –41.1 | –41.7 | –42.2 | –42.8 | –43.3 | –43.9 | –44.4 | –45.0 |
| –50 | –45.6 | –46.1 | –46.7 | –47.2 | –47.8 | –48.3 | –48.9 | –49.4 | –50.0 | –50.6 |
| –60 | –51.1 | –51.7 | –52.2 | –52.8 | –53.3 | –53.9 | –54.4 | –55.0 | –55.6 | –56.1 |

### Celsius (Centigrade) to Fahrenheit

**The vertical column on the left indicates 10 C intervals; 1 C intervals are shown across the top.**

| Degrees Centigrade | Degrees Fahrenheit | | | | | | | | | |
|---|---|---|---|---|---|---|---|---|---|---|
| | 0 | 1 | 2 | 3 | 4 | 5 | 6 | 7 | 8 | 9 |
| 40 | 104.0 | 105.8 | 107.6 | 109.4 | 111.2 | 113.0 | 114.8 | 116.6 | 118.4 | 120.2 |
| 30 | 86.0 | 87.8 | 89.6 | 91.4 | 93.2 | 95.0 | 96.8 | 98.6 | 100.4 | 102.2 |
| 20 | 68.0 | 69.8 | 71.6 | 73.4 | 75.2 | 77.0 | 78.8 | 80.6 | 82.4 | 84.2 |
| 10 | 50.0 | 51.8 | 53.6 | 55.4 | 57.2 | 59.0 | 60.8 | 62.6 | 64.4 | 66.2 |
| + 0 | 32.0 | 33.8 | 35.6 | 37.4 | 39.2 | 41.0 | 42.8 | 44.6 | 46.4 | 48.2 |
| – 0 | 32.0 | 30.2 | 28.4 | 26.6 | 24.8 | 23.0 | 21.2 | 19.4 | 17.6 | 15.8 |
| –10 | 14.0 | 12.2 | 10.4 | 8.6 | 6.8 | 5.0 | 3.2 | 1.4 | –0.4 | –2.2 |
| –20 | –4.0 | –5.8 | –7.6 | –9.4 | –11.2 | –13.0 | –14.8 | –16.6 | –18.4 | –20.2 |
| –30 | –22.0 | –23.8 | –25.6 | –27.4 | –29.2 | –31.0 | –32.8 | –34.6 | –36.4 | –38.2 |
| –40 | –40.0 | –41.8 | –43.6 | –45.4 | –47.2 | –49.0 | –50.8 | –52.6 | –54.4 | –56.2 |
| –50 | –58.0 | –59.8 | –61.6 | –63.4 | –65.2 | –67.0 | –68.8 | –70.6 | –72.4 | –74.2 |

*12. Air Pressure*

1 standard atmosphere (sea level) = 1013.2 millibars = 29.92 inches of mercury = 760 millimeters of mercury = 14.7 pounds per square inch.

### Conversion of Inches to Millibars
The Vertical Column on the Left shows Whole Inches; Tenths of Inches are Across the Top. For Equivalents Not Shown, Multiply Inches By 33.86.

| | | | | | Millibars | | | | | |
|---|---|---|---|---|---|---|---|---|---|---|
| Inches | 0 | 0.1 | 0.2 | 0.3 | 0.4 | 0.5 | 0.6 | 0.7 | 0.8 | 0.9 |
| 25 | 847 | 850 | 853 | 857 | 860 | 864 | 867 | 870 | 874 | 877 |
| 26 | 880 | 884 | 887 | 891 | 894 | 897 | 901 | 904 | 908 | 911 |
| 27 | 914 | 918 | 921 | 924 | 928 | 931 | 935 | 938 | 941 | 945 |
| 28 | 948 | 952 | 955 | 958 | 962 | 965 | 968 | 972 | 975 | 979 |
| 29 | 982 | 985 | 989 | 992 | 996 | 999 | 1002 | 1006 | 1009 | 1013 |
| 30 | 1016 | 1019 | 1023 | 1026 | 1030 | 1033 | 1036 | 1040 | 1043 | 1046 |
| 31 | 1050 | 1053 | 1057 | 1060 | 1063 | 1067 | 1070 | 1074 | 1077 | 1080 |

### Conversion of Millibars to Inches
The Vertical Column on the Left are Millibars in Multiples of 30; Across the Top, Millibars Are in Multiples of 5. for Equivalents Not Shown, Multiply Millibars by 0.0295.

| | | | Inches | | | |
|---|---|---|---|---|---|---|
| Millibars | 0 | 5 | 10 | 15 | 20 | 25 |
| 860 | 25.40 | 25.54 | 25.69 | 25.84 | 25.99 | 26.13 |
| 890 | 26.28 | 26.43 | 26.58 | 26.72 | 26.87 | 27.02 |
| 920 | 27.17 | 27.32 | 27.46 | 27.61 | 27.76 | 27.91 |
| 950 | 28.05 | 28.20 | 28.35 | 28.50 | 28.64 | 28.79 |
| 980 | 28.94 | 29.09 | 29.23 | 29.38 | 29.53 | 29.68 |
| 1010 | 29.83 | 29.97 | 30.12 | 30.27 | 30.42 | 30.56 |
| 1040 | 30.71 | 30.86 | 31.01 | 31.15 | 31.30 | 31.45 |

## SUGGESTED REFERENCE

AMIRAN, D. H. K., and SCHICK, A. P., *Geographical Conversion Tables*. Zurich, Switzerland: International Geographical Union, Aschmann and Scheller, A. G., 1961.

# Index

207